Urban Deer Havens

Urban Deer Havens

Clark E. Adams
and
Cassandra LaFleur Villarreal

CRC Press
Taylor & Francis Group
Boca Raton London New York

CRC Press is an imprint of the
Taylor & Francis Group, an **informa** business

First edition published 2020
by CRC Press
6000 Broken Sound Parkway NW, Suite 300, Boca Raton, FL 33487-2742

and by CRC Press
2 Park Square, Milton Park, Abingdon, Oxon, OX14 4RN

CRC Press is an imprint of Taylor & Francis Group, LLC

Library of Congress Cataloging-in-Publication Data
Library of Congress Control Number: 2020935366

ISBN: 978-0-367-48162-9 (hbk)
ISBN: 978-1-138-30507-6 (pbk)
ISBN: 978-1-003-03832-0 (ebk)

Typeset in Times LT STD
by Deanta Global Publishing Services, Chennai, India

This book is dedicated to the loving memory of Dr. Clark E. Adams, whose dedication to the study of urban wildlife will never be forgotten. Not only was he an outstanding professor, but he was also a committed and loving human being.

Contents

Preface

This book was developed based on a working hypothesis that human invasions, i.e., urban sprawl, into the geographic range of deer will eventually require some form of management intervention by the invaders. This supposition has been supported in both the professional and popular literature on urban deer. It was written because we were unable to find one publication that addressed all the pertinent issues related to urban deer. Information was scattered throughout various books, journal articles, conference proceedings, government documents, websites, data sets, and among the anecdotal tales of our colleagues. The results from prior surveys of selected respondents from state departments of natural resources (DNRs) provided evidence of the magnitude of urban deer management problems throughout the United States.

We begin by defining terms in order to get all readers on the same page regarding our use of these terms throughout the book. One example is all terms related to "urban" and another is "deer." For example, this book consists of a thorough examination of selected cervid (deer) species that are known to inhabit urban communities in the United States. These deer species include white-tailed deer (*Odocoileus virginianus*), Key deer (*O. v. clavium*), moose (*Alces alces*), elk (*Cervus elaphus*), mule (*Odocoileus hemionus*), and black-tailed deer (*O. h. columbianus*). This book is the first attempt to examine the similarities and differences in those factors that allow the selected cervids to exist and thrive in urban habitats. This information has never been collected, collated, reviewed, and published in one document. Yet, all five are known to inhabit urban communities within their geographic ranges. This book is intended to provide a more balanced approach, including what research has been done and needs to be done to more fully understand the cervid niche in urban environments.

This book is the first comprehensive source of information on urban deer management that includes: (1) national distributions and estimated cervid populations within states, (2) their ecological role as a keystone species, (3) management activities that promote or preclude their urban existence, (4) available funding and public education options, and (5) conflict resolutions, e.g., humans with deer and humans over deer. An important consideration is the examination of those factors that promote or preclude successful conclusions of urban deer management programs' objectives.

It has become evident that the challenges and opportunities related to urban deer have been noticed in the wildlife profession. As a result, this book is a long-overdue tool for teaching and learning about urban deer management. Based on the magnitude of impacts deer have on urban communities, this book will be a valuable technical and educational asset to many organizations, groups, and individuals engaged in some level of urban deer management. The obvious examples include city, state, and federal governments; college and university students pursuing urban wildlife management degree programs; urban residents who desire to become more informed about deer management in their neighborhoods; public and private animal damage control specialists; wildlife management, conservation, and ecology societies; and professional wildlife biologists.

Specific content provides a more holistic examination of urban cervids including (1) the similarities and differences of their ecology and environmental impacts, (2) management strategies and human dimensions considerations concerning urban cervids in general, and (3) the use of specific examples. Comprehensive tables review (1) urban community deer management options and plans, (2) examples of states and cities engaged in urban deer management efforts for selected cervids, and (3) estimated numbers of the five selected cervids in each state. National distribution maps of deer diseases and deer/vehicle accidents will allow readers to assess these risks in their own states.

This first edition of *Urban Deer Havens* is designed to address "the rest of the story," including those issues that are usually left out of manuscripts that focus primarily on how to alleviate the problems associated with urban white-tailed deer. For example, this book provides a basic framework of information that gives the reader an understanding of factors that promote or prevent the presence of deer in urban communities. The lack of information concerning several important examples of urban cervids in conjunction with a proliferation of information on white-tailed deer only is an incomplete and biased presentation.

Acknowledgments

Writing and completing this book was far more difficult than I could have ever imagined. First, I would like to thank my husband Gilbert for his unlimited encouragement, support, and patience through this process. Second, I would like to thank my parents, Thomas and Carmen LaFleur, for the countless pictures of urban deer that they sent me, and in particular my father for all his advice and editorial contributions. I would also like to thank our editor, Randy Brehm, for her compassion and assistance while completing *Urban Deer Havens*. I know how difficult completing this book was for her too. Finally, I would like to thank the Adams family for their immense support throughout the whole process.

Cassandra LaFleur Villarreal

Author Bios

Clark E. Adams was an Emeritus Professor in the Department of Wildlife and Fisheries Sciences (WFSC) at Texas A&M University in College Station, Texas. He received a BS in biology from Concordia Teachers College, Seward, Nebraska; an MS in biology from the University of Oregon; and a PhD in zoology from the University of Nebraska–Lincoln. He chaired the Conservation Education Committee for The Wildlife Society (TWS), edited the newsletter for the Human Dimensions of Wildlife Study Group, was a member of the Urban Wildlife Management Working Group, and has chaired many committees for the Texas Chapter of TWS. He was a past president of the Texas Chapter of TWS and TWS Southwest Section. Since 1981, he and his students have conducted and published many national, regional, and statewide studies on the public's activities, attitudes, expectations, and knowledge concerning wildlife. He developed the degree option in urban wildlife and fisheries management for the WFSC and developed and taught the senior-level urban wildlife management course. He was the senior author on two other books titled Texas Rattlesnake Roundups and Urban Wildlife Management, third edition. He was the recipient of the 2015 Outstanding Achievement in Urban Wildlife Conservation award from the TWS Urban Wildlife Working Group. Adams completed a 51-year teaching and research career on August 31, 2015.

Cassandra LaFleur Villarreal has worked as an Environmental Specialist in the Dallas/Fort Worth area for over five years specializing in water quality. She received her BS in wildlife and fisheries sciences from Texas A&M University in 2014. Cassandra developed an interest in urban wildlife after taking an Urban Wildlife Management course in 2012. In 2013, she began working for Dr. Clark E. Adams as a student worker. While working for Dr. Adams, Cassandra contributed to Urban Wildlife Management, third edition. She developed an alternative method of examining urban biodiversity using literature (peer-reviewed articles and field guides), rehabilitation center intake data, national data sets on bird species per state, and mapping methods such as Global Biodiversity Information Facility (GBIF). She conducted a meta-analysis of the relationship between a species resident by state(s) and observation of that species in an urban setting. This analysis included all native species of terrestrial herpetofauna, birds, and mammals known to exist in the United States. Her analysis and results led to the production of an entire new section and five new chapters in the third edition.

1 Urban Deer Introduction

As humans encroach into rural areas, deer have become urban animals.

Bob Kornegay 2016

This book was developed based on a working hypothesis that human invasions, i.e., urban sprawl, into the geographic range of deer will eventually require some form of management intervention by the invaders. A logical extension of this hypothesis would be that urban communities – throughout the United States – have a real, emerging, or potential urban deer management problem. Unfortunately, there are not enough individuals trained in deer ecology, theory, and general practice of urban deer management when compared to the number of urban communities requiring this type of professional expertise. Additionally, the rules of engagement are vastly different in urban areas when compared to rural habitats. This supposition was supported by both the professional and popular literature on urban deer. For example, we were unable to find one publication, of any kind and at any level of treatment, which addressed all the pertinent issues related to urban deer. Information was scattered throughout various books, journal articles, conference proceedings, government documents, PDF documents and data sets, and the anecdotal tales of our colleagues. Finally, results from prior surveys of selected respondents from state departments of natural resources (DNRs) and universities provided evidence of the near total lack of preparation by past and upcoming wildlife biologists regarding urban deer management problems throughout the United States.

This book is *not* just about the management issues related to urban deer, i.e., animal damage control. In order for deer to effectively exist in human built environments and for us to successfully negotiate human/deer agendas, we need to understand a broad base of information regarding our coexistence. The contents of this book were designed to provide a more holistic examination of human/deer interactions in urban environments. The goals for writing this book were fairly straightforward: (1) to compile a body of information that stimulates the reader's curiosity about the urban deer living in their neighborhoods, and present it in a way that would be accessible for most readers; (2) to expand the reader's knowledge about how natural and urban ecosystems work; (3) to challenge readers to examine the role their personal actions may have in creating at least some of the urban deer issues covered in this book; and (4) to give readers an opportunity to apply their new knowledge and understanding through personal actions that promote sustainable approaches to urban deer management.

SIDEBAR 1.1 DEER ARE A MIXED BLESSING

Deer represent both a benefit and a problem for humans. The benefits can be tangible. For example, national incomes from retail sales, taxes, wages, and job opportunities from 10.9 million deer hunters totaled $12.9 billion in 2011 (Fuller 2016). Non-consumptive tangible benefits included observing, feeding, or photographing wildlife living in the neighborhood, or taking a special wildlife-watching trip. Bird watching was the primary focus for 16.3 million residents away from home, and for 45.1 million residents around their homes. Observing large land mammals (including, but not exclusively, deer) was reported by 11.8 million residents away from home and 19.7 million residents around their homes. Total expenditures for wildlife viewing including trips, wildlife-viewing equipment, transportation, food and lodging, educational materials, camping equipment, and other expenses were $75.9 billion. Specific expenditures to observe deer were not reported (U.S. Department of the Interior 2016).

A report (Northern Economics, Inc. 2006) pertaining only to Alaska moose provided mid-2005 estimates of consumptive and non-consumptive use values. Consumptive use values from moose harvest, hunting experience, subsistence values, and meat salvage were $38.3 million. Non-consumptive use values from tourism activities tied to moose viewing and photography totaled $3.9 million.

On the other hand, urban deer can also cause detrimental effects due to deer-vehicle collisions, disease transmission (e.g., Lyme disease), landscape/garden damage, habitat degradation, declining deer herd health, and public safety. These problems are particularly apparent when urban deer become "overabundant." Overabundance is usually translated in terms of deer/acre that exceed (1) biological or (2) cultural carrying capacities. Levels of habitat degradation determine the first indicator of overabundance. Unacceptable levels of detrimental effects on urban residents determine the second indicator of overabundance. As will be discussed in later chapters, benefits and detrimental impacts play major counterproductive roles in expediting urban deer management programs. In general, urban

FIGURE 1.1 White-tailed deer fawn demonstrates the "Bambi syndrome" (courtesy Larry Reis, FLIKR creative commons).

FIGURE 1.2 A white-tailed deer fawn feeding on tomato plants in a backyard (courtesy Glane, FLIKR creative commons).

residents have a "love/hate" relationship with deer, and can develop strong emotional attachments, e.g., the Bambi syndrome – a fairy tale sense of reality concerning deer (Figure 1.1), or "rats with horns" (Figure 1.2).

1.1 TERMINOLOGY

The first step in becoming more informed on the complex issues related to human/deer interactions in urban settings is to understand the language of the discipline, i.e., relevant terminology. To begin, some terms that are used repeatedly throughout this book are *manage*, *wildlife*, and the urban context including the following: *urban*, *urbanization*, *urban ecology*, *urbanized areas* and *clusters*, and *urban sprawl*.

Manage means to indirectly or directly attempt to take charge of another species' existence. Wildlife management can be defined as the manipulation of wildlife populations and habitat to achieve a goal (Sargent and Carter 1999). There are usually four goals: increase populations, decrease populations, harvest for sustained yield, and monitor the population, or in some cases, do nothing (Krausman et al. 2013).

Wildlife includes amphibians, reptiles, birds, and mammals living in the wild or "free-ranging, undomesticated animals in natural habitats" (Krausman et al. 2013). The Wildlife Society defines wildlife as "free-living animals of major significance to man" (Krausman 2002). The Natural Resources Defense Council (2006) defines wildlife as "animals living in the wilderness without human intervention." *Urban wildlife* includes those species that have found habitat in towns and cities (Krausman et al. 2013).

The wildlife focus of this book is on white-tailed (*Odocoileus virginianus*) and Key deer (*O. v. clavium*), moose (*Alces alces*), elk (*Cervus elaphus*), mule (*Odocoileus hemionus*), and black-tailed deer (*O. h. columbianus*). As large native game (except Key deer) and endangered (Key deer) species, this assemblage of cervids presents unique management challenges in urban habitats.

The presence of various species of wildlife found in urban ecosystems was predicated on the characteristics of the "urban wildlife syndrome, UWS" (Pickett et al. 2011, as cited by Adams 2016). The UWS, when applied to deer living in urban ecosystems, suggested that deer have certain characteristics, conditions, or behaviors, including:

1. Generalists with high reproductive capacity (i.e., *r* strategists): Deer can utilize urban resources at will and produce more twin and triplet offspring when compared to rural populations.
2. High dispersal abilities: Deer move easily from one urban green zone to another, and in and out of town at will.
3. High rates of innovation: Deer are urban exploiters using every alternative resource available for survival. Many wooded suburban environments, such as parks, greenbelts, golf courses, and roadsides, meet the needs of deer.
4. High level of risk-taking: Deer cross streets and highways somewhat oblivious of car traffic. Also, they may attempt to

jump over fences that are too high and sometimes become impaled.

5. Rapid life histories: Does can be impregnated within six months of life and produce multiple fawns.
6. Biotic homogenization: "Consequently, as cities expand across the planet, biological homogenization increases because the same 'urban-adaptable' species (e.g., deer, author emphasis) become increasingly widespread and locally abundant in cities across the planet" (McKinney 2006).
7. Elevated densities associated with human subsidies: Urban communities provide deer with all of the resources required to maintain higher reproductive and survival rates when compared to deer in rural habitats.
8. Reduced wariness: Predator/prey concerns have been reduced or are nonexistent. Deer-vehicle collisions are common.
9. Increased intraspecific aggression: As urban deer density increases, they vigorously compete with one another over limited resources, e.g., food, space, or mates.
10. Increased tolerance of human presence: Deer can become the pets of urban residents (Figure 1.3). Consistently observed in close proximity to humans in parks, in urban neighborhoods, on golf courses, and downtown. The fight/flight distance between deer and urban residents is reduced or nonexistent. For example, elk may habituate to human activity if the disturbance is predictable and harmless (Thompson and Henderson 1998).

FIGURE 1.3 Urban residents may attempt to domesticate wild deer as pets. The deer in this picture was "rescued" as a fawn and raised to an adult (courtesy Clark E. Adams).

1.2 BUREAU OF CENSUS DEFINITION OF *URBAN*

Urban, in this context, includes both cities and suburbs. Also known as the "built environment," the urban landscape includes places where most of the property is devoted to all things man-made and/or maintained: buildings of all shapes and sizes, manicured lawns and landscaped office parks, cemeteries and vacant lots, strip malls and parking structures, elementary schools and college campuses, and airports and warehouse districts (Figures 1.4 and 1.5). A substantial portion of this land is covered with some kind of impervious surface in the form of either structures or pavement. The plant life is often native to other parts of the world or highly hybridized, and thus requires much caretaking in the form of mowing, pruning, weeding, watering, herbicides, pesticides, and fertilizers. *Urban* can also be characterized as "large concentrations of people and industrial activity that consume more available energy and material than can be produced, and which produce more wastes than can be assimilated with the relatively small area they occupy" (Rees 1997).

Urbanized areas and *urban clusters* now account for 80.7% of the U.S. population (Geographic Products Branch 2010, Figure 1.6). An urbanized area (N = 486 nationwide) can be defined as "a large central place and adjacent densely settled census blocks that together have a

FIGURE 1.4 Typical suburban landscaped lawn demonstrating habitat simplification (courtesy Clark E. Adams).

FIGURE 1.5 Typical metropolitan areas with high-rise buildings, concrete surfaces, roads, and automobile traffic (courtesy John Davis).

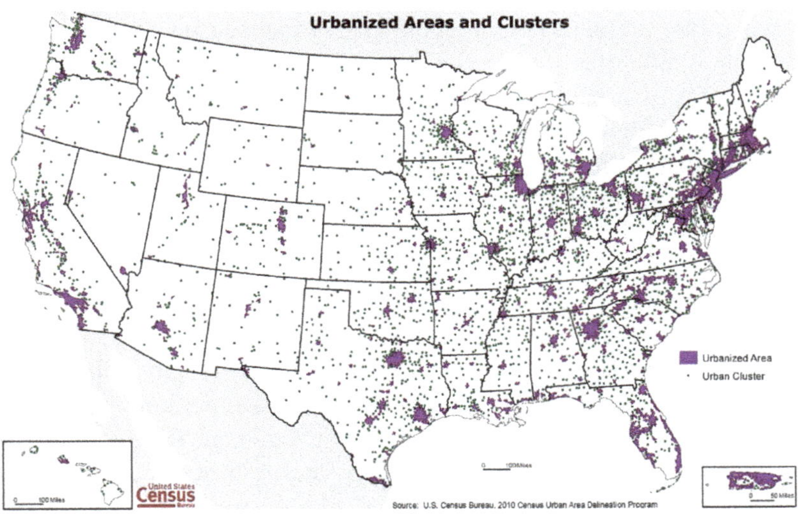

FIGURE 1.6 Extent of urbanized areas (purple) and urban clusters (green dots) in the United States in 2010 (Geographic Products Branch 2010).

total population of 50,000 or more people." Urban clusters (N = 3,087 nationwide) consist of "at least 2,500 and less than 50,000 people" (Geographic Products Branch 2010). The United States has a total land area of nearly 2.3 billion acres. Of this, 3% (61 million acres) is classified as urban land, representing an increase of 1 million acres since 2002 (Nickerson et al. 2011). The U.S. urban population increased by 12.1% from 2000 to 2010 to 73.6%, outpacing the overall growth rate of 9.7% during the same period. For example, there are 34 communities in Texas classified as urban, and 84.7% (22.7 million) of the Texas population reside in these urban communities (Geographic Products Branch, U.S Census Bureau 2010). Over 5% of the surface area of the United States is classified as urban and associated developed areas. In the next 25 years, U.S. urbanized areas will increase to 9.2% of the total land base (McKinney 2006).

1.3 ECOLOGICAL DEFINITION OF *URBAN*

Ecology is the study of relationships between the biotic and abiotic components of an ecosystem. *Urban ecology* is "the study of humans in cities, of nature in cities, and the coupled relationship between humans and natural systems" (Marzluff et al. 2008). The traditional scientific definition is used for studies of the distribution and abundance of organisms in and around cities, and on the biogeochemical budgets of urban areas. For urban planners, urban ecology has focused on designing the environmental amenities of cities for people, and on reducing environmental impacts of urban regions (Pickett et al. 2011). The bottom line is that urban ecology research must include any and all anthropomorphic influences within a human-dominated ecological system. The overarching goal is to understand how humans and ecological processes can coexist and help urban societies to become more sustainable (Marzluff et al. 2008).

Urban sprawl refers to the post-WWII demographic shifts that led to "the widespread expansion of metropolitan areas through buildings and shopping centers farther and farther from urban centers and lacing them together with more and more major highways, usually without planning" (Wright and Boorse 2011). Urban sprawl set the stage for human ecosystem conflicts that had not existed before, and reinforced a separation between people and much of the natural world. Also, urban sprawl fragments, isolates, and degrades natural habitats; simplifies and homogenizes species composition; disrupts hydrological systems; and modifies energy flow and nutrient cycling (Alberti and Marzluff 2004, citing several authors).

The process of urban sprawl begins and is perpetuated by roads. Roads are the primary mechanism used to begin and perpetuate access to wildlands. The urban–rural continuum is typically described as six different land zones beginning with a natural area (Zone T1) and ending with the urban core (Zone T6) (Figure 1.7). Zones T3 to T6 are generally those

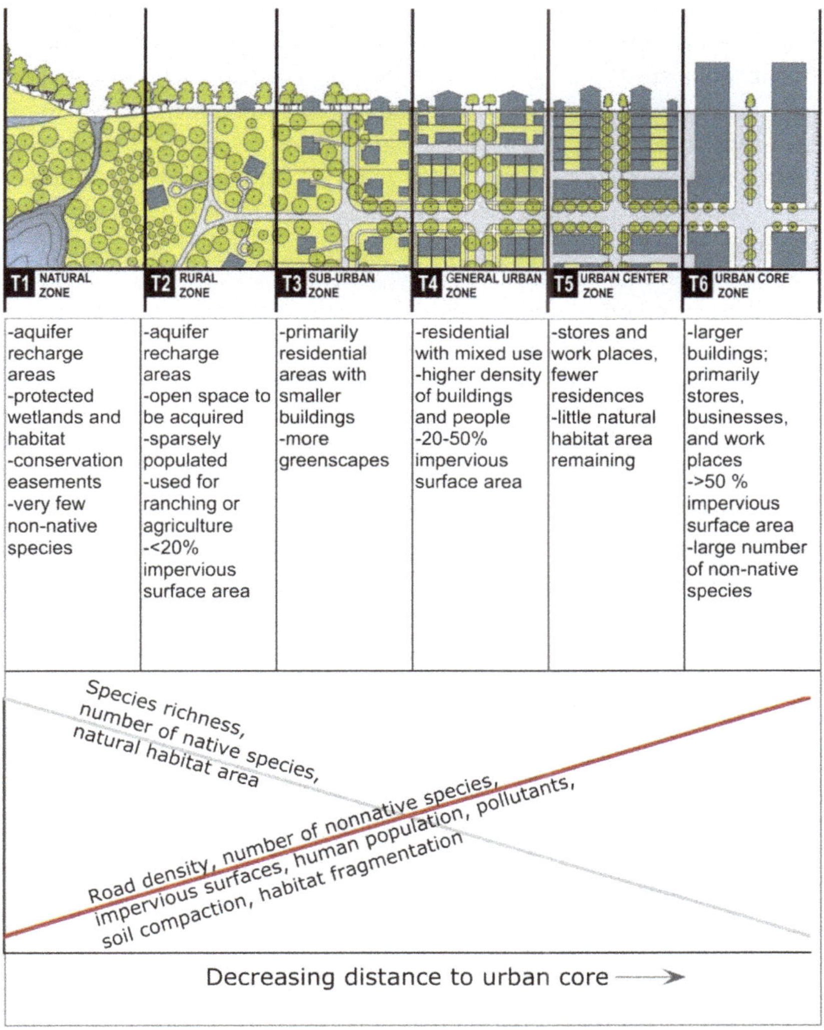

FIGURE 1.7 The urban–rural continuum with the six representative land zones. As distance to the urban core decreases, abiotic and biotic processes are affected in various ways (Dorney 1977, McKinney 2002, Pickett et al. 2011). Rural–urban transect provided by Duany Plater-Zyberk & Company (courtesy Michaela Murphy, Adams 2016 Figure 3.4).

included in the definition of an urban area or matrix. Each zone is characterized by its cultural history, abiotic characteristics, and biotic features. For example, as the distance between Zones T1 and T6 decreases, there is a steady decline in the species richness, number of native species, and natural habitat (McKinney 2008). On the other hand, there is a steady increase in road density, number of nonnative species, impervious surface covers, human population, pollutants, soil loss and compaction, and habitat fragmentation. An alternative description of the urban matrix is one that "consists of a large number (diversity) of habitats associated with vacant lots, wasteland areas, small patches of remnant habitats, riparian areas along streams, ditches, and canals, and a diverse and large group of residential habitats" (Nilon 2010, page 47, citing earlier studies). In themselves, residential areas can also present a diversity of habitat types including:

1. Parks, cemeteries, open spaces.
2. Woodlands, closed forests residential.
3. Residential streets, yards, and fencerow trees.
4. Residential lawns and shrubs.
5. Scarcely vegetated residential areas (Nilon 2010).

Urban sprawl causes extensive habitat changes, resulting from different levels of human influence, i.e., invasion, on the natural habitat (Figure 1.7). These changes force a selective process on urban deer populations, making them urban exploiters. As such, humans have decided that urban deer require their assistance, i.e., management, to live or die in habitat-altered urban areas.

Of note is the lack of comprehensive and integrated research on the biological scale of impacts of urban development on *the deer* (Polfus and Krausman 2012, Table 2. Ungulate response to development along the continuum of the biological scale of impacts described by Johnson and St-Laurent (2011)).

Biological scale of impacts includes:

1. Individual behavioral responses: Movement away from disturbance, flight response, increased vigilance, altered foraging rates, changes in maternal activities, avoidance of development, and habituation.
2. Individual physiological: Increased heart rate, respiration, and stress hormone concentration.
3. Population responses: Prolonged and severe physiological or behavioral changes can alter vital rates and be detected at higher biological scales.

4. Ecological community responses: Proximity to human development may also alter interspecific relationships (i.e., predation, competition) and, thus, influence the ecological community composition and distribution.

1.4 DEFINITION OF DEER

Deer (cervids) are hoofed (ungulates) grazing or browsing mammals, with branched bony antlers that are shed annually and typically borne only by the male. The term *urban deer* has been used to describe deer that have adapted readily to urban and suburban habitats throughout many portions of North America (McAninch 1995).

1.4.1 Taxonomy of Urban Deer

We start this section by asking readers to recall the last time they engaged in the process of classification or attempting to bring order to a disorganized condition. For example, imagine you are given a box containing 2,972 different kinds of vertebrate animals that needed to be sorted given their similarities and differences. Believe it or not, some human minds cannot tolerate disorganization. In the field of zoology, these individuals are called "taxonomists" or "systematists." They make their living by bringing organization to the rich animal diversity that exists in our world. Some are "splitters," i.e., obsessed over subtle differences (e.g., use geographic range and morphological differences to develop *subspecies* divisions), and others are "lumpers," who have a more inclusive taxonomic approach. A species is often defined as the largest group of organisms in which two individuals can produce fertile offspring, typically by sexual reproduction. For example, the species name identifies those populations of deer that can interbreed with one another, e.g., white-tailed deer (*O. virginianus*) from Minnesota and Texas can breed, if given the opportunity, and produce more of the same species.

At this point we are going to use the science of "naming things," i.e., taxonomy, to organize the place of deer in the animal kingdom (Table 1.1). By using deer classification, we use special aspects of their anatomy that first make them a mammal, followed by those characteristics that set them apart from other species of mammals. In Table 1.1, the morphological and geographic diversity of cervids was demonstrated by an overwhelming array of subspecies names, e.g., white-tailed deer (N = 16), mule deer (N = 7), moose (N = 4), and elk (N = 4). We chose to include only Key deer as one of the urban white-tailed deer subspecies because of their unique morphology, geographic distribution, available literature (Table 1.2), and endangered species status.

TABLE 1.1
Classification of U.S. Urban Deer

Kingdom	Animalia (102,992)			
Phylum	Chordata (animals with backbones)			
Subphylum	Vertebrata (animals with a boney skeleton and skull, 2,972 species)			
Class	Mammalia (mammals, 428 species)			
Order	Artiodactyla (even-toed hoofed mammals = ungulates)			
Family	Cervidae (the deer family excluding caribou)			
Genus	*Odocoileus*	*Odocoileus*	*Alces*	*Cervus*
species	*virginianus*	*hemionus*	*alces*	*elaphus*
Common name	White-tailed deer[a] (WTD)	Mule deer[b] (MD)	Moose[c]	Elk[d]

Source: Integrated Taxonomic Information Systems, www.itis.gov
Note:

[a] Subspecies: (*O. v. clavius*, Key deer), (*O. v. leucurus*, Columbian WTD), (*O. v. ochrourus*, Rocky Mountain WTD), (*O. v. couesi*, Coues WTD), (*O. v. dacotensis*, Dakota WTD), (*O. v. texanus*, Texas WTD), (*O. v. macrourus*, Kansas WTD), (*O. v. Mcilhennyi*, Avery Island WTD), (*O. v. osceola*, Florida coastal WTD), (*O. v. seminolus*, Florida WTD), (*O. v. nigribaris*, Blackbeard Island WTD), (*O. v. hiltonensis*, Hilton Head Island WTD), (*O. v. venatorius*, Hunting Island WTD), (*O. v. taurinsulae*, Bulls Island WTD), (*O. v. virginianus*, Virginia WTD), and (*O.v. borealis*, woodland WTD).

[b] Subspecies: (*O. h. columbianus*, black-tailed deer), (*O. h. sitkensis*, Sitka black-tailed deer), (*O. h. californicus*, California MD), (*O. h. fuliginatus*, southern MD), (*O. h. peninsulae*, Peninsula MD), (*O. h. eremicus*, desert MD), and (*O. h. hemionus*, Rocky Mountain MD).

[c] Subspecies: (*A. a. americanus*, eastern moose), (*A. a. andersoni*, northwestern moose), (*A. a. gigas*, Alaska moose), and (*A. a. shirasi*, Shiras moose).

[d] Subspecies: (*C. c. roosevelti*, Rossevelt elk), (*C. c. nannodes*, Tule elk), (*C. c. manitobensis*, Manitoban elk), and (*C. c. nelson*, Rocky Mountain elk).

1.4.1.1 Deer Characteristics That Identify Them as Mammals

First, deer are considered mammals because they:

1. Produce milk in mammary glands to suckle fawns.
2. Have body hair.
3. Can maintain a constant internal body temperature despite the ambient temperature.
4. Produce offspring in a uterus, i.e., placental development.
5. Use their metabolic rate to control heat production, and sweat glands to help cool the body.
6. Have a four-chambered heart which completely separates oxygenated and unoxygenated blood.

TABLE 1.2
Sample of Urban Florida Key Deer Publications from 1922 to 2008

1. Barbour, T., and G.M. Allen. 1922. The white-tailed deer of the eastern United States. *Journal of Mammalogy* 3(2):65–78.
2. Barrett, M.A., P. Stiling, and R.R. Lopez. 2005. Long-term changes in plant communities influenced by Key deer herbivory. *Natural Areas Journal* 26:235–243.
3. Barrett, M.A., and P. Stiling. 2006. Key deer impacts on Hardwood Hammocks near urban areas. *Journal of Wildlife Management* 70(6):1574–1579.
4. Barrett, M.A., and P. Stiling. 2006. Effects of Key deer herbivory on forest communities in the Lower Florida Keys. *Biological Conservation* 129:100–108.
5. Braden, A.W., R. R. Lopez, C.W. Roberts, N.J. Silvy, C.B. Owen, and P.A. Frank. 2008. Florida Key deer (*Odocoileus virginianus clavium*) use and movements along a highway corridor. *Wildlife Biology* 14:155–163.
6. Carlson, P.C., G.T. Tanner, J.M. Wood, and S.R. Humphrey. 1993. Fire in Key deer habitat improves browse, prevents succession, and preserves endemic herbs. *Journal of Wildlife Management* 57:914–928.
7. Dickson, J.D., III. 1955. An ecological study of the Key deer. *Florida Game and Fresh Water Fish Commission Technical Bulletin* 3, Tallahassee, Florida, USA.
8. Folk, M.J., and W.D. Klimstra. 1991a. Reproductive performance of female Key deer. *Journal of Wildlife Management* 55:386–390.
9. Folk, M.J., and W.D. Klimstra. 1991b. Urbanization and domestication of the Key deer (*Odocoileus virginianus clavium*). *Florida Field Naturalist* 19:1–9.
10. Harveson, P.M., R.R. Lopez, N.J. Silvy, P.A. Frank. 2006. Comparison of survival estimates using age-specific mortality and radiotelemetry data for Florida Key deer. *Southeastern Naturalist* 5:149–156.
11. Harveson, P.M., W.E. Grant, R.R. Lopez, N.J. Silvy, and P.A. Frank. 2006. The role of dispersal in the metapopulation dynamics of Florida Key deer. *Ecological Modeling* 195:393–401.
12. Harveson, P.M., R.R. Lopez, B.A. Collier, N.J. Silvy. 2007. Impacts of urbanization on Florida Key deer behavior and population dynamics. *Biological Conservation* 134(3):321–331.
13. Harveson, P.M., R.R. Lopez, N.J. Silvy, and P.A. Frank. 2004. Source-sink dynamics of Florida Key Deer on Big Pine Key, Florida. *Journal of Wildlife Management* 68(4):909–915.
14. Klimstra, W.D. 1985. The Key deer. *Florida Naturalist* 58:2–5.
15. Klimstra, W.D., and A. Dooley. 1990. Foods of the Key deer. *Florida Scientist* 53:264–273.
16. Lopez, R.R., N. Silvy, B. Pierce, P. Frank, M. Wilson, and K. Burke. 2004. Population density of the endangered Florida Key Deer. *Journal of Wildlife Management*, 68(3), 570–575.
17. Lopez, R.R., N. Silvy, R. Labisky, and P. Frank. 2003. Hurricane impacts on Key Deer in the Florida Keys. *Journal of Wildlife Management*, 67(2), 280–288.

(*Continued*)

TABLE 1.2 (CONTINUED)
Sample of Urban Florida Key Deer Publications from 1922 to 2008

18. Lopez, R.R., N.J. Silvy, P.A. Frank, and B.L. Pierce. 2003. Effects of supplemental feeding on Key deer behavior in urban areas. *Florida Scientist* 66(4):267–272.
19. Lopez, R.R., N.J. Silvy, P.A. Frank, S.W. Whisenant, and D.A. Jones. 2003. Survival, mortality, and life expectancy of Florida Key deer. *Journal of Wildlife Management* 67(1):34–45.
20. Lopez, R.R., N.J. Silvy, R.N., Wilkins, P.A. Frank, M.J. Peterson, and M.N. Peterson. 2004. Habitat-use patterns of Florida Key Deer: Implications of urban development. *Journal of Wildlife Management* 68(4):900–908.
21. Lopez, R.R., P.M. Harveson, M.N. Peterson, N.J. Silvy, and P.A. Frank. 2005. From the field: Changes in ranges of Florida Key deer – does population density matter? *Wildlife Society Bulletin* 33(1):343–348.
22. Nettles V.F., C.F. Quist, R.R. Lopez, T.J. Wilmers, P. Frank, W. Roberts, S. Chitwood, and W.R. Davidson. 2002. Morbidity and mortality factors in key deer (*Odocoileus virginianus clavium*). *Journal of Wildlife Diseases* 38(4):685–692.
23. Parker, I.D., A.W. Braden, R.R. Lopez, N.J. Silvy, D.S. Davis, and C.B. Owen. 2008. Effects of US 1 Project on Florida Key Deer mortality. *Journal of Wildlife Management* 72(2):354–359.
24. Parker, I.D., D. Watts, R. Lopez, N. Silvy, D. Davis, R. McCleery, and P. Frank. 2008. Evaluation of the Efficacy of Florida Key Deer Translocations. The Journal of Wildlife Management, *72*(5), 1069–1075.
25. Peterson, M.N., R.R. Lopez, E.J. Laurent, P.A. Frank, N.J. Silvy, and J. Liu. 2005. Wildlife Loss through Domestication: the Case of Endangered Key Deer. Conservation Biology 19(3):939–944.
26. Peterson, M.N., R.R. Lopez, N.J. Silvy, C.B. Owen, P.A. Frank, and A.W. Braden. 2003. Evaluation of deer-exclusion grates in urban areas. Wildlife Society Bulletin 31(4):1198–1204.
27. Peterson, M.N., R.R. Lopez, P.A. Frank, B.A. Porter, and N.J. Silvy. 2004. Key deer fawn response to urbanization: is sustainable development possible? Wildlife Society Bulletin 32(2):493–499.
28. Peterson, M.N., R.R. Lopez, P.A. Frank, M.J. Peterson, and N.J. Silvy. 2003. Evaluating capture methods for urban white-tailed deer. Wildlife Society Bulletin 31(4):1176–1187.

    Quist, C.F., V.F. Nettles, S.J.B. Manning, D.G. Hall, J.K. Gaydos, T.J. Wilmers, and R.R. Lopez. 2002. Paratuberculosis in Key deer, (*Odocoileus virginianus clavium*). Journal of Wildlife Diseases 38:729–737.
29. Roberts, C.W., B.L. Pierce, A.W. Braden, R.R. Lopez, N.J. Silvy, P.A. Frank, and D. Ransom. 2006. Comparison of Camera and Road Survey Estimates for White-Tailed Deer. Journal of Wildlife Management 70(1):263–267.
30. Watts, D. E., I. D. Parker, R. R. Lopez, N. J. Silvy, and D. S. Davis. 2008. Distribution and abundance of endangered Florida Key deer on outer islands. Journal of Wildlife Management 72:360–366.

1.4.1.2 Deer Characteristics That Identify Them as Cervids

Second, cervids (deer) are a distinct subset (family Cervidae) within the class Mammalia because they:

1. Grow boney antlers, which are the defining characteristic of the family Cervidae.
2. Walk on hooves, which are highly modified hands, feet, fingers, toes, and nails.
3. Use a special method of digestion that involves a four-chambered rumen or paunch type stomach preventing the need to "chew their food thoroughly before swallowing." The cervid stomach has four chambers: the first stores food, and the other three digest it. Food is regurgitated, re-chewed (called chewing the "cud"), and swallowed again, entering sequentially the other three stomach chambers.
4. Are herbivores (plant eaters); most are browsers (eating leaves, shoots, soft vegetation, and twigs), but some are also grazers (eating mostly grass).
5. Use a hard palate and tongue, rather than teeth on the top of their mouth, for browsing.

1.4.1.3 Deer Characteristics That Identify Them as Different Species or Subspecies

All species of cervids have become urban exploiters by necessity or fortuitously due to human invasions of their natural habitat. The distinguishing features of those deer species (subspecies) commonly found in urban communities (Figure 1.8) include size, weight, antler configuration, statewide distribution (Table 1.3), and unique external appearance. Differences in size and antler configurations are the obvious characteristics that distinguish cervid species. The information sources were amazingly redundant or contradictory in some cases, e.g., hoof to shoulder height, weight, and

FIGURE 1.8 Common urban deer in the United States. Note size comparisons in relation to an urban resident (courtesy Cassandra LaFleur Villarreal and C.E. Adams).

TABLE 1.3
Resident Deer (Cervids) by State 2016

State	White-tailed Deer	Mule Deer	Moose	Elk	Key Deer
Alabama	■				
Alaska		■	■	■	
Arizona	■	■		■	
Arkansas	■			■	
California		■		■	
Colorado	■	■	■	■	
Connecticut	■		■		
Delaware	■				
Florida	■				■
Georgia	■				
Hawaii		■			
Idaho	■	■	■	■	
Illinois	■				
Indiana	■				
Iowa	■				
Kansas	■	■		■	
Kentucky	■				
Louisiana	■				
Maine	■		■		
Maryland	■				
Massachusetts	■		■		
Michigan	■		■	■	
Minnesota	■		■	■	
Mississippi	■				
Missouri	■				
Montana	■	■	■	■	
Nebraska	■	■		■	
Nevada		■		■	
New Hampshire	■		■		
New Jersey	■				
New Mexico	■	■		■	
New York	■		■		
North Carolina	■			■	
North Dakota	■	■	■	■	
Ohio	■				
Oklahoma	■	■		■	
Oregon	■	■	■	■	
Pennsylvania	■			■	

(*Continued*)

TABLE 1.3 (CONTINUED)
Resident Deer (Cervids) by State 2016

State	White-tailed Deer	Mule Deer	Moose	Elk	Key Deer
Rhode Island	■				
South Carolina	■				
South Dakota	■	■		■	
Tennessee	■				
Texas	■	■		■	
Utah		■	■	■	
Vermont	■		■		
Virginia	■			■	
Washington	■	■	■	■	
West Virginia	■			■	
Wisconsin	■		■	■	
Wyoming	■	■	■	■	
Number of States	**46**	**19**	**18**	**26**	**1**

FIGURE 1.9 Note diminutive size of an adult male Key deer (courtesy Bonnie Masdeu, FLIKR creative commons).

antler sizes. Often, it was impossible to determine the "original" source(s) of information.

The *Key deer*, also called the "toy deer," is the smallest subspecies of white-tailed deer (Figure 1.9). Bucks stand from 28 to 32 inches at the shoulder and weigh an average of 80 pounds. Does range from 24 to 28 inches at the shoulder and weigh an average of 65 pounds.

They live in the lower Florida Keys spread around over 20+ different islands, with the largest concentration (75%) in Big Pine and No Name Keys (see Figure 2.12 in the following chapter). This restricted geographic range in the Florida Keys isolated them from the mainland white-tails, which gave them the subspecies (*O. v. clavium*) and the EPA threatened classifications. Over-hunting and habitat loss due to urbanization nearly decimated the entire Key deer population, causing them to be listed as endangered (March 11, 1967) by the Endangered Species Act (Key deer, *O. v. clavium*). Key deer feed on 180 different species of plants, the main source being red and black mangrove, Indian mulberry, silver palm, and thatch palm. They cannot tolerate salt water in their diet. Territorial behavior is limited to a buck's defense of a receptive doe from other bucks, and is not used to defend a specific territory size or area (Klimstra et al. 1974). Mating is promiscuous during the rut, but is the only association males have with females; i.e., there is no male/female pair bonding.

The *white-tailed deer* (Figure 1.10) derive their name and distinguish themselves from other deer with their broad, 10 to 11 inches long tail (Figure 1.11). The tail flips up when threatened and as a sign of danger to other deer in the herd. The tines on their antlers grow off a single beam and are not forked. They stand from 2.6 to 3.3 feet at the shoulder. Average weights of adult males and female white-tails are 150 and 100 pounds, respectively (Dewey 2003). The white-tailed deer diet is extensive and diversified. Depending on the season and location, their diet may include growing tips of trees and shrubs, grass, clover, landscape plants, fruit, nuts, acorns, fungi, lichens, and garden crops. They have the most

FIGURE 1.10 White-tailed deer buck (courtesy Clark E. Adams).

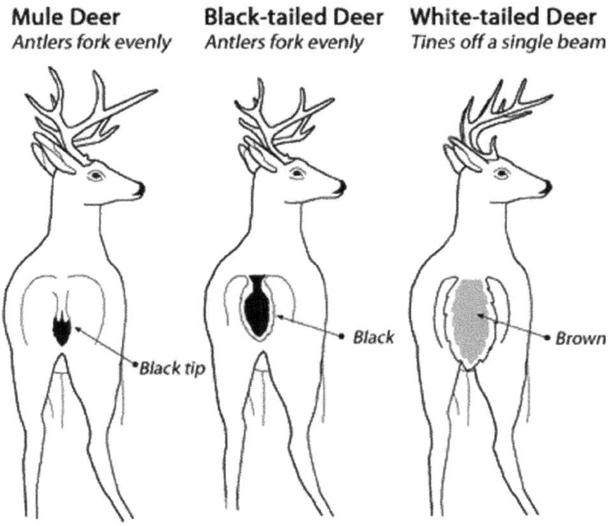

FIGURE 1.11 Antler and tail characteristics that distinguish mule and white-tailed deer. http://wdfw.wa.gov/living/deer.html (Washington Department of Fish and Wildlife, Accessed March 12, 2017).

extensive statewide range (N = 46) of any of the deer species (Table 1.3). Their geographic success is most likely a product of evolution, genetic superiority, wary behavior, adaptability, improved range (Deckman 2003), and urban sprawl. Both males and females are promiscuous; i.e., males mate with several females and females may mate with one or more males. Of note is that 12 states, namely Arkansas, Georgia, Illinois, Michigan, Mississippi, Nebraska, New Hampshire, Ohio, Oklahoma, Pennsylvania, South Carolina, and Wisconsin, have designated the elegant white-tailed deer as an official state symbol.

The *mule deer* (Figure 1.12) derive their name for their ears, which are large like those of a mule. They also distinguish themselves in tail appearances (see Figure 1.11). Their tail is 5 to 8 inches long and flips up when threatened to warn other deer in the herd. Mule deer are slightly larger than white-tailed deer. They stand from 3 to 3.5 feet at the shoulder. Adult black-tailed deer bucks weigh 140 to 200 pounds, and adult does weigh 90 to 130 pounds. Antlers fork evenly in both subspecies. The mule deer diet is as extensive as the white-tailed deer, and consists of many of the same categories of edible plant material. Promiscuous mating between males and females occurs during the rut. They are found in 19 states (Table 1.3). The mule deer is not a recognized state symbol in any state.

The *American elk* (Figure 1.13), or wapiti, is one of the larger cervids. They have the most restricted statewide range (N = 19, Table 1.3). Later discussions on the geographic range of elk illustrate how extensive

FIGURE 1.12 Urban mule deer in Helena, Montana.

FIGURE 1.13 An urban elk looking inside a restaurant (courtesy Fiona Sinclair, FLIKR creative commons).

they used to be distributed in America compared to their contemporary restricted range. Mature males weigh 700 to 800 pounds, and females can tip the scale at 500 to 550 pounds. Males stand 4.9 feet at the shoulder compared to 4.3 feet for females. A total of 159 forbs, 59 grasses, and 95 shrubs are listed as elk forage and categorized according to their relative value (Kufeld 1973). Elk eat grasses, sedges, dandelions, aster, hawkweed, violets, clover flowers, and other soft plant foods during summer months.

FIGURE 1.14 Urban moose (Free images, Public Domain, Courtesy Karen Laubenstein, USFWS).

In winter, favored woody plants include wintergreen, cedar, red maple, eastern hemlock, sumac, jack pine, staghorn, and basswood. A 700 pound bull elk will eat about 21 pounds of food a day. Elk live in large herds throughout the year. During the rut, cows and calves form harems with one or two mature bulls. Bugling (bellow escalating to squealing whistle ending with grunting) occurs in the fall. By bugling, the bull is advertising his fitness to cows, warning other bulls to stay away, or announcing his readiness to fight. Only bulls have long forked antlers up to 3.9 feet long and weigh up to 40 pounds when completely bone hard. Cows, calves, and yearlings live in loose herds or groups. Bulls live in bachelor groups or alone ("Elk Facts," n.d.). The elk is the official state symbol for Utah.

Moose (Figure 1.14) are the largest of all the deer species. They are found in 18 states (Table 1.3). Mature males can weigh as much as 1,600 pounds and stand up to 7 feet from hooves to shoulders. Adult females weigh between 440 and 790 pounds and are 5–5.9 feet tall from hooves to shoulders. Mature males grow enormous, distinctive, palm-shaped antlers that can be over 6 feet across and weigh 50 to 70 pounds. Moose are great swimmers and feed on aquatic vegetation, e.g., water lilies, duckweed, and bur weed which provide sodium in their diet. The principal staples of the moose diet are the leaves and twigs of the willow, followed by other woody browse species such as gooseberry and buffalo berry. Daily food consumption is 10–12 and 22–26 up to 73 pounds per day in the winter and summer, respectively. Moose are solitary animals, but during the rut season, moose will herd into groups for the purpose of finding mates. Cows will attract bulls with loud moans. Bulls need to seek out females in heat and defend breeding rights from other bulls by antler fighting. A healthy, well-fed cow moose will have twins as much as 75% of the time. Triplets are not uncommon. The moose is the official state symbol in Alaska and Maine.

All four of the larger cervids, e.g., white-tailed and mule deer, moose, and elk, are found in six states: Idaho, Montana, North Dakota, Oregon, Washington, and Wyoming (Table 1.3). Their presence in these states would suggest that urban communities could be dealing with all four in their deer management programs. One can imagine the differences in attempting to coexist with a white-tailed deer compared to a moose.

1.5 CONTEMPORARY RELEVANCE OF A TEXT ON URBAN DEER

Several statements, addressing the contemporary relevance of the development of the first book on urban deer, address its practical value in contemporary literature.

1. Nearly every urban community within a cervid's geographic range is now or will need to address how the residents desire to "manage" their coexistence. Yet there is no single and readily accessible compendium of information that addresses the important ecological issues that need to be considered in the management process.
2. Groups that might require a practical guide to urban deer management issues might include: (1) city, state, and federal governance units involved in urban cervid management; (2) university students pursuing urban wildlife management degree programs; (3) urban residents who desire to become more informed about cervid management in their neighborhoods; (4) public and private animal damage control specialists; and (5) professional wildlife biologists.
3. The information in this book is offered as a nontechnical practical guide that offers a comprehensive synthesis of professional, public, and popular literature on urban cervid population ecology and management. This exercise was considered absolutely necessary given the literature proliferation over the past three decades; the lack of public knowledge on how to access, understand, and apply this information; topic redundancy in published literature; and identifying missing information required to conduct urban deer management more effectively. Finally, we are profoundly humbled by and grateful to those professionals who provided the extensive body of information used to produce this book.

4. This book contains examples of and references to existing state urban deer management plans (nearly every state has one) including options and strategies, how to address and interpret the human dimensions or "people factor" issues, and the cautionary aspects of the management process, e.g., deer census results.
5. It may be difficult for urban communities with deer management problems to find and employ competent urban deer specialists to help them. Later, we discuss a national study that revealed a consistent deficit in state departments of natural resources' level of involvement in urban deer management and the number of university wildlife degree programs that focused on urban community wildlife management needs.

1.5.1 Urbanization: Collision of an Invasive Hyperkeystone Species with a Keystone

Urbanization is a spatial phenomenon by which towns and cities are formed and become larger with increasing concentrations of the population (McDonald and Marcotullio 2014). "Urbanization can be characterized as an increase in human habitation, coupled with increased per capita energy and resource consumption and extensive modification of the landscape, creating a system that does not depend principally on local natural resources to persist" (McDonnell and Pickett 1990). Their list of structural features, unique to urbanization includes dwellings, factories, office buildings, warehouses, roads, pipelines, power lines, railroads, channelized stream beds, reservoirs, sewage disposal facilities, landfills, and airports, among so many other unnatural inclusions. Urbanization is an important driver of climate change and pollution, altering both biotic and abiotic ecosystem properties within, surrounding, and even at great distances from urban areas (Grimm et al. 2008). Of note is the consistency of a habitat's structural and functional changes resulting from urbanization, regardless of geographic location, and always beginning with a road! In fact, "urban land-use change may be homogenizing the US, producing residential ecosystems/landscapes that are more similar to each other than to the natural ecosystems that they replace" (Groffman et al. 2014). A rough analogy would be that just as Walmart stores in any state are exactly the same in structure and function – so are urban communities.

1.5.1.1 Humans (*Homo sapiens*) Are the Invasive Species

Any discussion of urban deer needs to set the record straight regarding "invasive species." The term applies to an introduced species that can sustain self-replacing populations over several life cycles, produce reproductive offspring, and have the potential to spread over long distances, causing measurable detrimental economic, human health, and habitat effects. Did humans invade the deer's natural habitat, or did the deer invade the human's altered habitat? In other words, "who was there first?" Native deer species had but only three choices under such conditions: (1) adapt – become urban exploiters or adaptable species; (2) move – disperse or migrate; or (3) die – become threatened, endangered, or extinct. Adaptation or urban exploitation are more common than one would imagine. Adams (2016) and colleagues identified a total of 740 species of amphibians, reptiles, birds, and mammals that were observed in urban habitats. All five cervid species were included. It appears that urbanization is "setting the table" or providing an acceptable alternative to natural habitats for animals pre-adapted to the change. The human invasion of natural habitats follows a consistent, i.e., homogeneous, pattern of structural changes, as listed above and illustrated in Figure 1.7, regardless of the geographic location. So the invitation to deer to use the alternative resources is consistent and reliable.

1.5.1.2 Humans as a Hyperkeystone Species

A keystone species plays a major role in the survival (short- or long-term) of many other species in the ecosystem. There is evidence that prior to contemporary time, native people were keystone predators who once restructured entire ecosystems. They influenced the distribution, abundance, and behavior of wildlife, although they lacked any effective conservation strategy regarding their primary prey items (Jerozolimski and Peres 2003). When it comes to urban areas, Rees (1997) claims "humans are the creators of cities and therefore the keystone species in the urban system." Humans have the ability to alter both the abiotic and biotic conditions, and the structure and function of ecosystems, i.e., the need to change the natural to preferred (less objectionable) habitats. More specifically, human activities that support consideration as a keystone species are as follows (O'Neill and Kahn 2000, O'Neill 2001, Cunningham and Cunningham 2010).

1. Constructing both dispersal barriers and invasion pathways through habitat fragmentation.
2. Introducing other invasive and exotic deer including Axis deer (*Cervus axis*), Fallow deer (*Dama dama*), and Sika deer (*Cervus nippon*), and causing mass deer extinctions, e.g., Schomburgk's deer (*Cervus schomburgki*) and the Eastern elk (*Cervus canadensis canadensis*) as discussed below.
3. Changing ecosystem stability by altering environmental constraints and biota, e.g., ambient temperature; air, water, and land pollutions; and removal of natural predators.
4. Manipulating disturbance regimes by suppressing or increasing probabilities of occurrence, e.g., loss of forest and other natural ecosystems

to accommodate urban sprawl. Pre-settlement (1650) hardwood forests used to extend from the Atlantic coastline to the Mississippi River covering 80% of the total land mass. "Two centuries of logging and agricultural clearing were initiated that removed more than half of the forest cover and cut over almost all of the rest" (Thompson et al. 2013). This massive habitat change caused the extinction of the Eastern elk from the area (Table 1.3) prior to the Civil War.
5. Their effect on communities affects all trophic levels, e.g., producers and consumers. Humans replace natural vegetation with horticultural varieties for landscaping, which is a botanical bonanza for browsing deer.
6. Changing competitive relationships, e.g., predator/prey relationships and availability and abundance of food, water, and shelter alternatives.

The above *human only* impacts on all natural ecosystems are so pervasive and effective that they are now being examined as a hyperkeystone species (Worm and Paine 2016). "Humans have historically been considered as an inconsequential externality as if their effects belong in a separate category compared to other species and their interactions" (Worm and Paine 2016). For example, the existence of urban ecosystems is solely the product of human influences that modify (directly or indirectly) nearly every aspect of their abiotic and biotic surroundings regardless of habitat. Another unique characteristic of hyperkeystone species is their far-reaching impacts on multiple keystone species, including urban deer.

1.5.1.3 Deer as a Keystone Species

Keystone species have disproportionately large impacts on ecosystems. For example, the ecological impact of overabundant white-tailed deer in urban areas can affect (1) the distribution or abundance of many other species, (2) community structure by strongly modifying patterns of relative abundance among competing species, or (3) community structure by influencing the abundance of species at multiple trophic levels (Waller and Alverson 1997). The heavy and constant browsing of urban white-tailed deer can destroy entire vegetative communities (e.g., forest understory) upon which other species depend for their survival. In addition, browsing deer promote the selection of vegetative communities with less plant and animal biodiversity. These profound impacts on the ecology of urban ecosystems led Waller and Alverson (1997) to designate urban white-tailed deer as a "keystone" species. In comparison, the long-term cumulative effects of other larger introduced deer (e.g., moose) that become overpopulated and isolated, e.g., island populations, on the structure and function of native communities can be described as "wrecking balls" (McLaren et al. 2004).

CASE STUDY 1.1 REVIEWS OF PUBLISHED LITERATURE

Conducting a synthesis of available peer-reviewed published literature on urban deer proved to be difficult and frustrating, but tenacity and perseverance produced rewards. First, it is imperative that the right search term(s) be used which was not as obvious as one would think. Second, different search engines, e.g., AGRICOLA, BIOSIS, EBSCO, JSTOR, Web of Science, Academic Search Complete, and Google Scholar gave vastly different results. All search results for the expected outcome, i.e., journal articles on urban deer, delivered many instances of redundant and inconsistent results, or citations that had nothing to do with the search term. Nevertheless, literature reviews are an absolutely required *a priori* process in the research and management process. Literature reviews provide guidance in terms of the past and present accomplishments and future directions regarding specific information required to conduct urban deer research or management.

A small scale (N = 31) example of the results of a literature search on urban Key deer is illustrated in Table 1.2. The term "Key deer" had to be in the title or listed in article terminology to be selected during the search. Masters theses and doctoral dissertations were not included. These are not definitive results, but represented examples of some past and contemporary (21st century) Key deer research. Table 1.3 represents a literature review that would have to be conducted in order to determine what has been done in the past and needs to be done in the future. Publication titles addressed the habitat impacts on and by Key deer, e.g., 2, 3, 4, 6, 15, 17, 18, 21, 29; population dynamics, e.g., 8, 10, 11, 12, 13, 16, 19, 22, 31; urban impact issues, e.g., 5, 9, 12, 18, 20, 23, 25, 27; management, e.g., 14, 24, 26, 28, 30; general ecology, e.g., 7, 14; and taxonomy, e.g., 1.

Now, based on the above analysis of Key deer research publications, what might be the future research emphases? One of the most obvious omissions was research on the human dimensions (HD) aspects of Key deer management. HD research usually includes public attitudes, activities, expectations, and knowledge related to the management of selected wildlife species. Given the rapid urbanization of the Florida Keys, the endangered status of Key deer, and daily encounters with people (Figure 1.15), HD research might be a prudent inclusion in future research agendas. As discussed below, HD was a dominant research emphasis on white-tailed and mule deer, but not Key deer. However, there was a strong connection between past HD research and acceptance of alternative management options to reduce deer numbers in urban communities. This emphasis would be totally inappropriate in Key deer HD studies. Rather, the HD emphasis would be on the

FIGURE 1.15 Key deer easily habituate to people (courtesy Ben Lawson, FLIKR creative commons).

conservation and management of Key deer, an endangered species, within urban communities.

A more extensive literature review of urban deer research included 84 peer-reviewed journal articles published between 1980 and 2015 (Adams 2016, Case Study 1.1). The majority of selected articles were published between 1997 and 2011 (Figure 1.16). Study site locations for the articles included 24 states and Canada, and four articles encompassed study site locations across the United States (Table 1.4). The majority of articles were published in the *Wildlife Society Bulletin* (*WSB*) (59.5%, N = 50) and the *Journal of Wildlife Management* (*JWM*) (11.9%, N = 10). H. Kilpatrick produced the greatest number (N = 15) of urban deer articles.

Key search words included "population control" (lethal and nonlethal), "human perceptions," "conflict management," "spatial and temporal use," "diet," "behavior," "genetics," "techniques and methods," "disease," and "affecting other species." The most prevalent key words assigned included "population control" (22%, N = 31) and "human perception" (15%, N = 21; Table 1.5). Seven articles discussed contraception and/or public perceptions of that management control technique. The key word "disease" was assigned to three articles, while "diet" and "genetics" were each assigned to only one manuscript. Seven articles employed modeling or simulation techniques to predict spatial and habitat use, survival analysis, or deer-vehicle collisions. Additionally, seven articles discussed property damage as a result of urban deer.

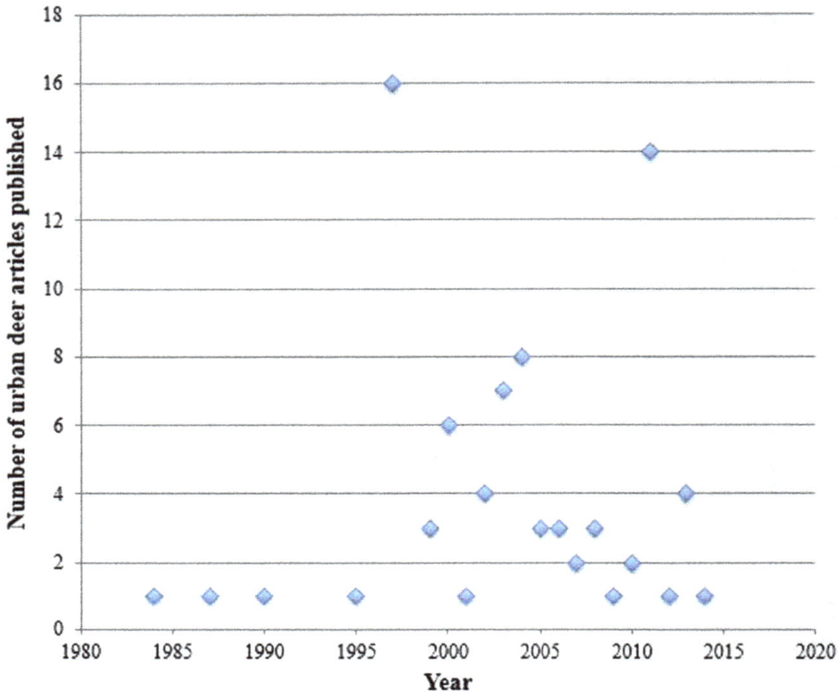

FIGURE 1.16 Years in which urban deer articles were published (Adams 2016, Case Study 1.1).

Since the 1980s urban deer research emphases were repetitive and focused primarily on population control and human perceptions (Table 1.5). There was a dominant focus on deer damage control. HD research may continue to be the focus of research and management for urban deer. Research omissions included disease transmission by deer in urban areas, how urban deer populations affect other plant and animal species, diet differences between urban and rural areas, and genetic information of deer in urban areas.

More articles focused on lethal when compared to nonlethal urban deer management options. The public has generally preferred nonlethal over lethal management strategies (Messmer et al. 1997a, Stout et al. 1997, Lauber and Knuth 2000, Bowker et al. 2003, Kilpatrick et al. 2007). On the other hand, state wildlife agencies perceived hunting and lethal methods as acceptable techniques for managing urban deer (Messmer et al. 1997b). Public acceptance of lethal management strategies depended on the method and the degree to which they were experiencing problems (Lauber et al. 2004, Stewart 2011). However, only two articles considered the perspective of state wildlife agencies. The public may have views on

TABLE 1.4
States Where Urban Deer Research Was Conducted

Location	Number of Articles
Across the United States	4
Canada	3
Colorado	2
Connecticut	15
Florida	6
Georgia	2
Illinois	6
Indiana	2
Iowa	4
Maine	1
Maryland	2
Michigan	1
Minnesota	5
Missouri	3
Nebraska	1
New Jersey	4
New York	12
Ohio	1
Oregon	1
Pennsylvania	3
South Carolina	5
Tennessee	1
Texas	4
Unknown	2
Utah	2
Washington	1
Wisconsin	1

Note: Some studies conducted research in multiple locations.

what deer control method they prefer, but actions that state agency biologists may implement are limited by legal, economic, and political motivations (Messmer et al. 1997a, Urbanek et al. 2011).

With increasing rates of urbanization and an expanding wildland-urban interface, urban deer populations will continue to increase. Therefore, research on the management and population control (Table 1.5) of urban deer will probably continue to be the dominant emphasis.

TABLE 1.5
Key Words Assigned To Urban Deer Articles

Key Word	Number of Articles Assigned To
Population control	31
Human perception	21
Lethal management	19
Spatial and temporal	18
Nonlethal management	15
Techniques and methods	13
Conflict management	10
Behavior	4
Disease	3
Affecting other species	3
Diet	1
Genetics	1

Note: Each article was assigned at least one key word based on article content.

The number of peer-reviewed manuscripts on urban white-tailed deer research in the *WSB* has grown steadily since 1997. One entire issue – 50 articles – of the *WSB* (1997: 25 (2)) was dedicated to urban white-tailed deer research. Later, the *WSB* devoted an entire issue (2011: 35 (2), 25 articles) to the ecology and management of white-tailed deer in developed landscapes. In all of these publications was the notable absence of urban deer research on mule deer, elk, and moose.

Even the news media are increasing their attention to urban deer herds. The cover story of the December 9, 2013 issue of *Time* was devoted to urban wildlife under the title of "America's pest problem" (Von Drehle 2013). VanDruff et al. (1996) explained the complex nature of conducting field research in urban settings, with an emphasis on habitat structure in urban compared to rural environments, and the problems resulting from frequent contact with humans during the research process. These issues are even more prevalent two decades later!

A prolific source of information on urban deer can be found on websites containing reviews of multiple sources and topics, e.g., "Resource Library" (Community Deer Advisor) and "A Guide to Making an Urban Deer Management Plan." We were confronted with a huge challenge to filter all of this information into a cohesive, coherent, and organized presentation focused on urban deer. This was particularly problematic in the development of Tables 3.1 to 3.3 as will be discussed later.

There were about 50 books published on white-tailed deer from 1962 to 2011. These books covered the natural history and ecology of, and how to hunt, white-tailed deer. An examination of books on urban deer revealed two recent examples including *The Urban Deer Complex* by A.J. DeRosa et al. (2014) and *Urban Deer and Nontraditional Hunting: A Thoughtful Guide for Not-Yet Hunters* by Ken Slaughter (2015). The subject matter of these two titles focused on urban deer management through bow hunting. There were no books on urban deer *per se*. Books on white-tailed deer, mule deer, moose, and elk emphasized hunting, trophies, cooking venison, or the aesthetics of deer in the wild. There were five books written on Key deer biology, ecology, conservation, and management. These results were somewhat disappointing, but provided a significant stimulus to pursue our production of this book on *Urban Deer Management*.

2 Urban Neighborhoods Guarantee Deer Occupation

It's as if an entire species has been spoiled by the comforts of civilization.

Bob Shaw 2011

The suburbs are attractive to deer for some of the same reasons they are attractive to people. There are natural areas, greenways, and parks that provide bedding areas, escape cover, and birth sites. Homes are landscaped with trees, shrubs, and herbaceous cover, which are appetizing and nutritious to deer. Wild and domestic predators have been extirpated or controlled. Deer are afforded the same conveniences and protection as suburban residents.

(Pennsylvania Game Commission 2007)

Deer are of special significance in urban and suburban America, regardless of their zoogeographic distributions. They share a common legacy in terms of (1) those factors that contributed to their prominence in urban America; (2) the human response to their presence; (3) ecological impacts on their habitats; (4) human health and safety issues; and (5) feasible and acceptable management strategies.

There is a great deal of information on the issues pertaining to overabundant deer in urban communities. This chapter has attempted to reduce the complexities of the issues by highlighting the principal components involved when deer become urban exploiters. All deer have demonstrated their capabilities of utilizing the alternative urban resources to their advantage, causing environmental damage and danger to human health and safety in urban environments. It is important for the urban residents to be exposed to the fundamental ecological, economic, cultural, and political issues that need to be considered with any species that exceeds the carrying capacity of its environment. White-tailed deer, in particular, are charismatic personifications of this problem.

2.1 FACTORS THAT CONTRIBUTED TO DEER ABUNDANCE IN URBAN AMERICA

Urban deer populations in North America represent one of the most challenging problems facing wildlife managers today and in the future. Deer are superbly adapted to exploit the resources in urban areas (Figure 2.1), and some deer populations may demonstrate temporary exponential growth patterns caused by:

1. Low abundance of natural predators. Large predators are the first species eliminated during urban sprawl.
2. Healthy and safe breeding habitat conditions, including patchy and defined edge habitats for deer.
3. Tolerance of urban disturbances, including human presence and their activities, e.g., golfing, jogging, bike-riding, and washing the car, among other activities.
4. Longer life spans in the city when compared to the country.
5. Lack of hunting in urban areas and regulatory protections at the state and/or federal levels.
6. High reproduction and survival rates in offspring.
7. Abundant alternative food resources in the form of ornamental shrubs, garden plants, succulent grasses, small plants, and supplemental feed. Enhanced urban landscapes provide year-around stable living conditions for deer when compared to fluctuations in forage availability on natural ranges.

FIGURE 2.1 White-tailed deer grazing on a street in Austin's Northwest Hills (photo © micklpickl/Flickr through a Creative Commons license).

In fact, the best thing that has ever happened to deer was the arrival of humans! With adequate cover and available food, deer successfully navigate sidewalks, traffic, and backyard fences, appearing quite comfortable with daily interactions involving humans, dogs, and vehicles ("Urban Deer Population – What You Should Know").

2.2 COUNTING DEER

2.2.1 Census Techniques

> Researchers and managers covet accurate, precise, and relatively inexpensive techniques for estimating population size.
>
> **Charles A. DeYoung 2011**

The urban deer management process begins by counting how many occupy the neighborhood. The count results need to be more accurate than a lot, a few, or none. Census precision requires real numbers, e.g., total population estimate based on count results. Census data is the only way, other than casual observations, to determine whether the deer population is increasing, decreasing, or stable. Management decisions on how many deer need to be captured and removed from the community and deer density rely on count data. The criteria required to achieve a "true" population inventory are impossible to achieve. For example, the implied assumptions for an unbiased deer count would be that:

1. All deer have an equal possibility of being counted.
2. All observers have equal opportunity to see deer.
3. The number of deer seen equals the number present.
4. The population is closed (no immigration or emigration).

An additional expectation is that census routes are the same during each sampling period and the same trained people are conducting the counts. The phrase "on the horns of a dilemma" has real meaning with *any* census technique used to estimate the size of a deer population. The dilemma is realized when one considers the variables that influence the population estimate results derived from the count procedure. Variables would include time of year or habitat type (amount of understory vegetation influences deer detection probabilities), capabilities of observers to recognize deer presence, time period between census before and after deer harvest (do deer move into or out of the community?), and possible mixing of deer herds along census lines.

A particularly cautionary review of animal population estimates pointed out the frequent use of convenience samples and the use of raw counts as indices of relative abundance (Anderson 2001). Convenience sampling occurs when surveys are conducted in easy to access areas, e.g., roads, utility corridors, parking areas, or areas of known but artificially high population densities, which are not representative of an animal's habitat locations. Index values are the result of raw counts that rely on detection probabilities that are assumed to be constant, regardless of habitat, season, animal detectability, and observer qualification variables. Anderson's (2001) comments on sampling suggest that census procedures follow the fundamental principles of sampling methods.

1. Determine how a sample is selected for a given population that does not violate the principles of randomness when compared to convenience sampling.
2. Utilize multiple repetitions of the sampling process in order to provide a more statistically reliable representation of the total population. Or to put it another way, "*repetitio est mater studiorum*" which is Latin for "repetition is the mother of learning."
3. The sampling method is important in determining how closely it represents the feasibility of access to the entire population.

Deer census results can, with great caution, be used to determine the required number of deer that need to be removed. The deer census process can also be a means for citizens to become a functional part of the deer management process. In some states, regardless of process and results, the deer census is required only so communities can get a permit from the state DNR to capture the deer. According to some urban wildlife biologists, the count may not have any relevance in developing the community deer management program. Instead, the plan is based on economic losses, health (deer and human) issues, and cultural carrying capacity which runs the gamut of "live and let live" to "kill them all."

2.2.2 The Census Process

The types of information expected from a credible census effort would include:

(1) How many deer counters were involved and trained to conduct the census?
(2) How were the census routes and distances chosen, i.e., where deer are known to be present or randomly chosen routes?
(3) What was the estimated total number of deer habitat acres in the urban community – including all green spaces?
(4) How was the value of (3) determined?
(5) How many acres of deer habitat were included in the census?

(6) How was the value of (5) determined?
(7) What was the estimated deer/acre, deer/mi^2, or deer/km^2 density based on census data?
(8) What was the total estimated deer population in the community based on census data?
(9) How did the census data determine how many deer should be harvested?

All census methods provide estimates only! Also, there are no fail-safe methods of estimating population size. In fact, there is no way of determining the exact number of urban deer anywhere and at any time. This would not be the case if all the urban deer could be run through a turnstile and counted individually. The wildlife management techniques manual (Pierce et al. 2012) provides extensive reviews of the many methods of estimating animal abundance. Some of the census methods that have been used in urban areas include:

1. Direct counts: Direct observations of deer along a specified line at night using a vehicle is considered the most consistent census method and is the best method for measuring deer density.
2. Aerial surveys: Use of helicopters, airplanes, or drones to count. Best for those managing large blocks of land. It does not work well in heavily forested areas because of limited visibility, unless conducted in the winter.
3. Trap/mark/re-sight: A subset of the deer population is marked with collars or ear tags. Sequential census periods examine the ratio of marked and unmarked deer – program MARK is required to manipulate data and calculate deer numbers and densities (White and Burnham 1999, McClintock et al. 2008, Grimm et al. 2014).
4. Distance sampling: This methodology is concentrated on detection functions which model the probability of detecting an animal, given its distance from the transect line. The key to distance sampling analyses is to fit a detection function to the observed distances, and use this fitted function to estimate the proportion of deer missed by the survey. Computer software, e.g., program Distance, is required to manipulate data and calculate deer numbers, detection probabilities, and densities (Thomas et al. 2010).

Census methods are similar in urban and rural habitats. All census procedures were originally developed to count rural deer herds. Urban census procedures are adaptations of the original rural methods. Most of the assumptions (discussed above and below) of rural census methods are also legitimate concerns in urban areas. Census techniques have become complicated in terms of statistical manipulations of raw data, e.g., programs MARK and Distance. This provision is referred to as "statistically-robust estimates of deer abundance" (Cobb 2014). It is beyond the scope and purpose of this book to explain the statistical intricacies of the programs MARK and Distance. The reader is directed to references that are cited, deer population biologists, biometricians, or faculty and students in wildlife management programs at state land grant universities to perform this function.

Cost and habitat relevance are obvious factors determining which method can be used in urban communities. "Choice of the most precise, cost effective method will be site specific and rely on information collected from a pilot study" (Amos et al. 2014, p. 362). The census should be conducted several days in sequence to produce an average (minimal) deer count. The total deer habitat sampled needs to be determined in order to calculate deer density. The definition of habitat that can be included in density calculations can be problematic. For example, total deer habitat is difficult to determine because urban deer utilize many of the structural features (Figure 2.2) resulting from urbanization in addition to natural and

FIGURE 2.2 White-tailed deer utilizing a garage.

built green spaces. Two sets of information result from these census procedures which are required to estimate deer abundance. Sampled deer/acre will result from dividing the average deer count by the sampled acreage. This average density value is multiplied by the total deer habitat in the urban community producing an estimate of the resident deer population per acre. Deer density/mi^2 results from dividing the total estimated deer population by 640 acres.

Two, less statistically robust, census methods, e.g., direct count and Lincoln–Peterson Index, are described below. From a conceptual and procedural context, they are easy to understand and expedite. It is also possible for the community residents to conduct their own deer census using the following procedures.

2.2.3 Direct Count

The census process begins by establishing zones of contact with the local deer population using a plat map of an urban community (Figure 2.3). Note that six different deer census zones have been laid out on the map. The census zones are approximately the same size and have at least one major road going through them. DNR wildlife biologists can provide training for "deer counters" among residents that reside in each zone. Being "deer counters" offers some residents the opportunity to become active participants in the deer management process, but as the first stage in the census process, it carries a heavy responsibility.

For example, all other aspects of the census process will rely on doing a thorough job when counting deer. The suggested procedure for counting deer is as follows:

1. All counters within each zone should conduct their deer census at the same time of the day; this would be usually early morning or evening. This will prevent them from counting the same deer herd that is moving around the community. Multiple counts are required to obtain averages in every zone sampled.
2. The census should be conducted in approximately the same location(s) within each zone. Multiple counters can divide up their zone for this purpose.
3. Counting can be done through direct observation, or spotlight detection along designated and consistent census lines. These are two cost-effective and uncomplicated census approaches.
4. The census should be repeated for about two weeks during the winter and summer (when the number of newborn fawns will be visible). The census data should include the number of bucks, does, and fawns observed by the deer counter.

FIGURE 2.3 Plat map of an urban subdivision partitioned into six census zones (courtesy Clark E. Adams).

5. There should be a designated data repository for the number of deer (bucks, does, and fawns) spotted in each zone, day, and season. This process requires proficiency in spreadsheet data entry and the ability to interpret and report the deer population dynamics from the data. Assistance from DNR deer biologists may be available.

If the above procedures are conducted with consistent diligence by all counters and multiple times, the result should provide a determination of relative deer numbers and density within urban neighborhoods. An additional benefit from conducting the census within specified zones may reveal clumped deer populations which might indicate supplemental feeding activities by some residents or an abundance of deer fodder, e.g., vegetable gardens or orchards, in that zone. It is reasonable to expect that the number of observations of deer within each zone will be a function of the actual numbers in that zone. For example, observations will be less frequent in a low deer density and more frequent in a high deer density. Another way of understanding this is that either deer will be hard to find, or there will be deer everywhere you look. Both conditions call for different urban deer management plans (as discussed in Chapter 3).

Next, a professional trapper, with experience and DNR regulatory clearances, should be hired to remove the excess number of deer determined from the credible estimate (options discussed in Chapter 3). The usual parameters considered to determine the "ideal" deer density in urban communities are prior determinations of biological and cultural carrying capacities discussed in Chapter 3. According to Campbell (2014), Hansen and Beringer (1997) suggested a density of 15 deer/km^2, and Tymkiw (2010) suggested <20 deer/km^2. Other studies found deer densities between 35 and 91 deer/km^2 (Kilpatrick et al. 2001, Bowman 2011). A more recent study reported an urban deer density of 62.2–65.8 deer/km^2 (Campbell 2014).

The recommended deer density values given frequently in popular (state urban deer management plans) and published literature were 15–20 deer/mi^2 or km^2. Given that a square mile equals 640 acres and a square kilometer equals 247 acres, we were unsure which metric would be accurate and acceptable. Personal communication with colleagues revealed an unwillingness on their part to suggest a recommended deer density by any metric. The bottom line is that each community has to decide how many deer per acre it and the deer can tolerate (as discussed in Chapter 3). The effectiveness of the management plan may be monitored in terms of reduced plant damage, deer-vehicle collisions, loss of deer from disease and starvation (the deer herd will appear healthier), and human/deer conflicts. However, expect annual additions to the deer herd from prolific reproduction and possible (urban deer are largely closed populations) immigrations (Figures 2.4 and 2.5).

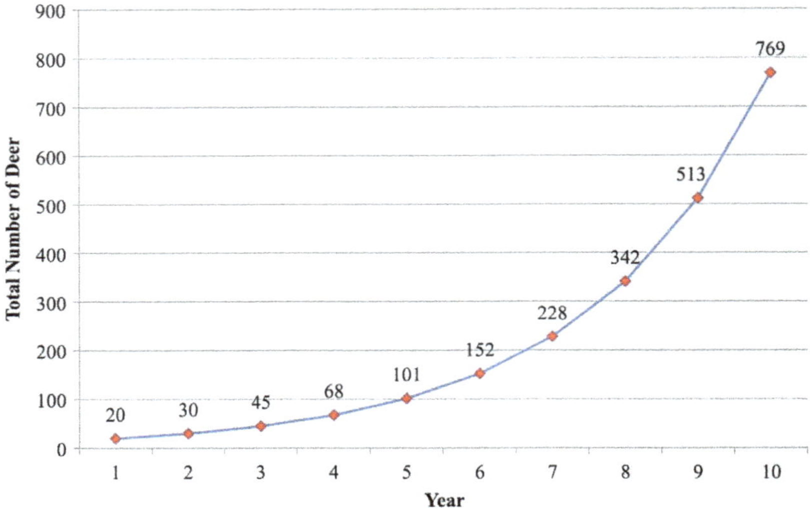

FIGURE 2.4 Example of exponential deer population growth. A deer population starting with 10 males and 10 females is projected for 10 years, assuming 50% fawn recruitment and 50% fawn mortality. Given none leave or others come into the population, a population beginning with 20 individuals will grow to 769, degrading and exhausting habitat resources.

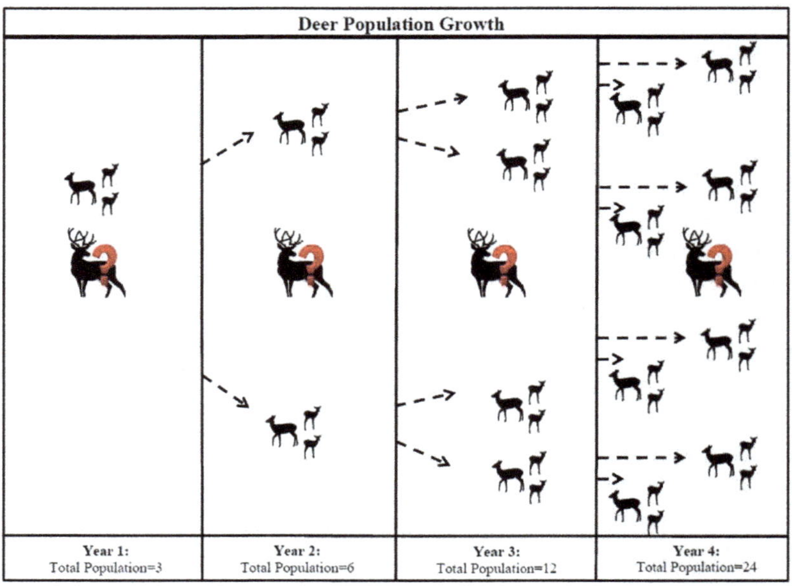

FIGURE 2.5 Schematic presentation of potential white-tailed deer population growth over a four-year period with twins – a common reproductive result in urban habitats. Of note is the low or unknown number of bucks required to produce the annual reproductive results (courtesy Cassandra LaFleur Villarreal).

In the absence of predation or hunting, this kind of reproduction can result in a deer herd doubling its size in one year. This fact was illustrated on the 1,146-acre George Reserve in southern Michigan where biologists at the University of Michigan have been studying the deer population since 1928. The deer herd grew from 6 deer in 1928 to 162 deer by 1933. In 1975, the George Reserve herd grew from 10 deer to 212 deer in five years (McCullough 1979 and 1984). Figure 2.5 illustrates how few bucks are required to inseminate does during each reproductive cycle. This buck/doe dynamic emphasizes the need to focus management efforts on female, not male, deer. The bottom line is that communities need to accept the fact that their plan will be an ongoing obligation as long as there are female deer in the neighborhood.

2.2.3.1 Application of Direct Count as a Population Estimate Procedure

The average number of deer counted during three consecutive nights using the same census lines, procedures, and observers= 500.

Total community deer habitat = 3,250 acres.

Portion of community deer habitat included in deer surveys = 325 acres.

Average deer/acre based on survey data = 500 deer/325 acres = 1.5deer/acre.

Total deer in community = 1.5 deer/acre × 3,250 acres = 4,875 deer.

Deer density/mi^2 = 4,875 deer/640 acres = 7.6 deer/mi^2.

2.2.3.2 Application of Trap/Mark/Re-sight as a Population Estimate Procedure

Trap/Mark/Re-sight (TMR) is an expensive option given the need to capture and mark a subset of the urban deer population with ear tags (Figure 2.6). Private companies will trap and mark subsets of the deer herd in preparation for future census activity. The logic behind the TMR census process is that the ratio of marked and unmarked deer observed during subsequent surveys can provide a population estimate. The TMR procedure is a replicate of the Lincoln-Peterson Index (LPI). The index formula is relatively simple:

Lincoln-Peterson Index

$$N = M \times S / R$$

N = population size estimate

M = marked individuals released

S = size of second sample

R = marked animals re-sighted in subsequent surveys

FIGURE 2.6 Urban deer with ear tags and collar for population estimate and home range analysis (courtesy Kara Campbell).

The fundamental assumptions of LPI used in TMR are as follows:

1. The population is closed so that N is constant.
2. Marked and unmarked animals are equally likely to be observed in each sample.
3. Sex and age classes are equally likely to be observed in each sample.
4. Capturing and marking do not affect observability.
5. Each census sample is random.
6. Marks are not lost between sampling occasions.
7. Marked animals mix randomly with unmarked animals.
8. Several census lines are required to provide an opportunity to re-sight marked deer that are distributed throughout the community.

Given the above draconian restrictions on LPI and TMR census procedures, we have attempted to demonstrate how field data can be applied to the LPI estimator given prior trap and mark procedures on an assumed closed population of deer. The population estimate is determined by the number of deer observed during each census sample which includes marked and unmarked deer, divided by the number of marked animals observed during each census sample. The number of marked deer is a constant (assuming no losses of markers or marked animals occur) during each census. Assume that during one census M = 458 (constant), S = 730, and R = 225.

$$N = 458 \times 730 / 225 = 1{,}486 \text{ deer}$$

The total community deer habitat = 3,250 acres.
The portion of community deer habitat included in deer surveys = 325 acres.
The average deer/acre based on survey data = 1,486 deer/325 acres = 4.6 deer/acre.
The total deer in the community = 4.6 deer/acre × 3250 acres = 14,950 deer.
The deer density/mi² = 14,950 deer/640 acres = 23.4 deer/mi².

The above example attempted to replicate some of the values from a study conducted in Fair Oaks Ranch, Texas (Campbell 2014). In this study, 458 deer (M) were captured and tagged (Figure 2.6). The population estimate was based on nine surveys, and the field data analyzed in program MARK suggested a total population estimate of 1,485 deer.

2.2.3.3 Application of Distance Sampling as a Population Estimate Procedure

Central to the concept of distance sampling is the *detection function*. This is the probability of detecting an object, given that it is at any distance *y* from the random line or point. In the example below, the distance *y* refers to the perpendicular distance from the transect line to a deer observation. Generally, the detection probability decreases (Figure 2.7) with the increase in the observation distance (Thomas et al. 2010). Apparently, the detection probability value is generated by program Distance using the array of distance values and the number of deer observed at each distance.

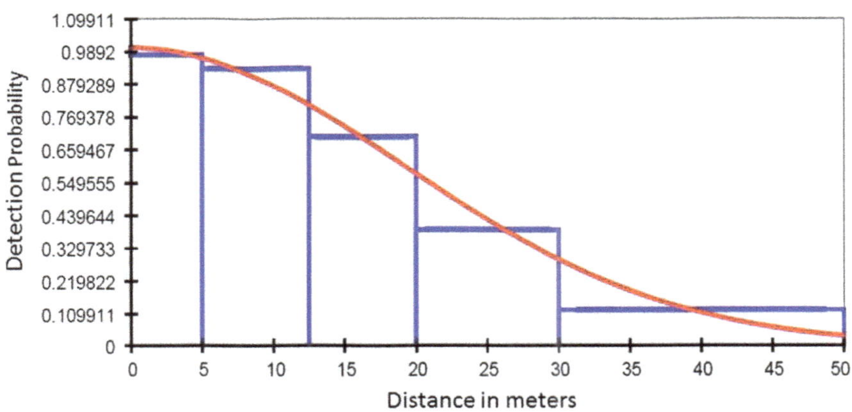

FIGURE 2.7 An example of a detection probability plot as generated by the analytical program Distance (Thomas et al. 2010). In this example, the probability of detection has dropped to 50% between 20 and 25 m distance from the observer.

Distance methods are designed to produce reliable estimates of abundance, regardless of observability, provided that the assumptions are met (Buckland et al. 1993). The assumptions of distance sampling are:

1. Objects on the line or point are detected with certainty.
2. Objects are detected at their initial location.
3. Measurements are exact as measured with a laser range finder.
4. Transect lines are positioned randomly within the study area.

The example of distance sampling (Figure 2.8) illustrates the kind of information required and obtained during the process. Twilight is the suggested time of day to conduct distance sampling surveys. Spotlight counts are used after dark. The most widely used form of distance sampling is line transect sampling. Transect lines can include city streets since urban deer are not startled by traffic and can be easily observed. Transect lines do not have to be straight, but can follow street designs. The majority of observations will occur close to the transect line (street) producing the "distance

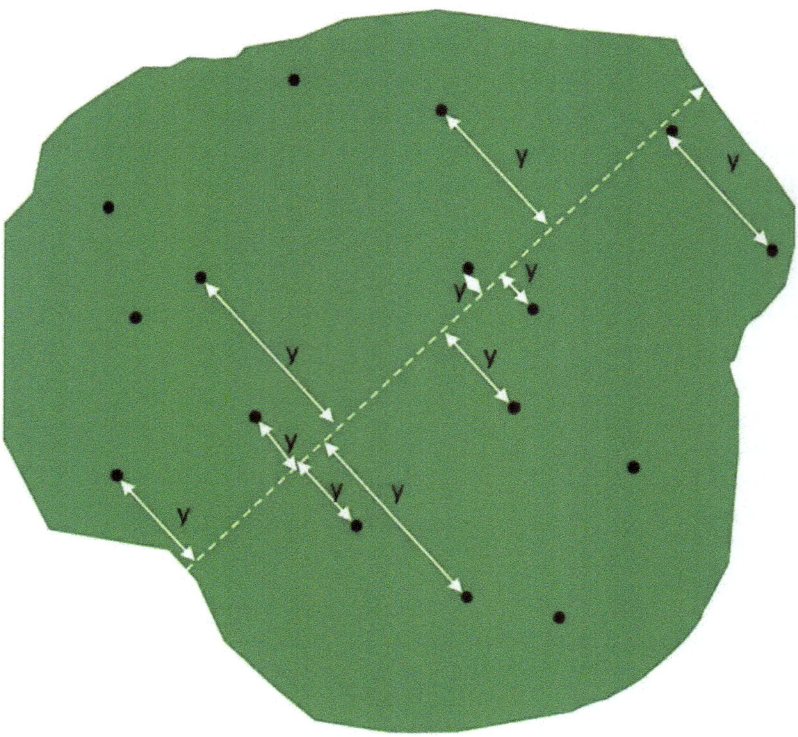

FIGURE 2.8 Distance sampling using several observation points along a transect line in a habitat of known (acres) size and transect lines of known (y) distances.

function" or "detection probability" characteristic of the distance sampling model (Figure 2.9). Spiked data might also arise if animals are attracted toward the observer. It is important that detections are made before any responsive movement occurs.

As a footnote to population estimate methods, our review of the literature left us with the conclusion that they are tools in theory with marginal practical and reliable applications. In fact, the dominant theme in contemporary writings on population estimation statistics focuses on rejecting the inherent biases in a selected method, often followed by an alternative theoretical approach, e.g., the inclusion of missing observations (MacKenzie et al. 2002) and inclusion of unmarked animals (Clement et al. 2015). However, "fixing the formula" is an ongoing necessity to provide greater accuracy in the census process.

2.2.4 Harvest Recommendations Based on Census Data

There are no known guidelines that provide harvest recommendations based on census data. All options are on the table. First, deer census results and resulting population size and density estimates assume an equally distributed population. Harvest decisions need to include differences in deer

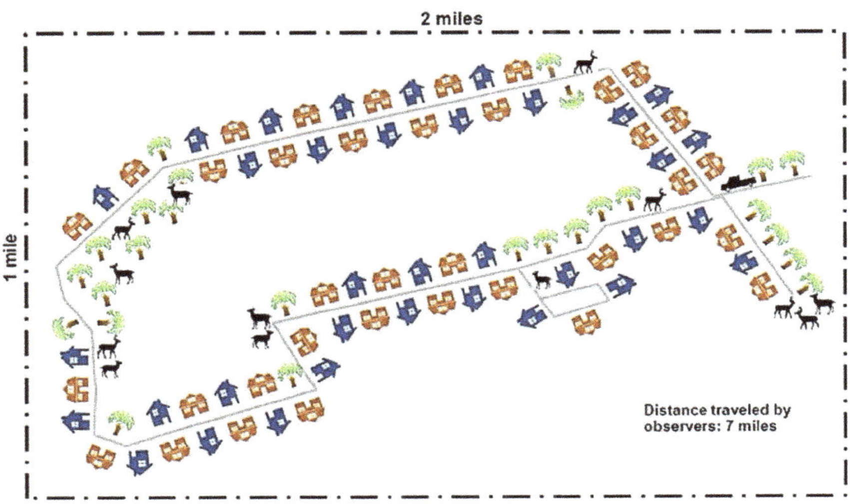

Total urban deer habitat included survey sample	2 mi²
Distance traveled by observers	7 mi
Average distance of animals seen from the road	0.009 mi (50 ft)
Total area surveyed	7 mi x 0.009 mi = 0.063 mi²
Total number of deer observed	13
Therefore, in 1 mi²... 13/0.063 x 2	413 deer
The estimated deer population is 413.	

FIGURE 2.9 Example of a road strip count of deer in an urban community.

densities within census zones in order to attack the root cause of an overpopulation problem. Decisions regarding where to harvest and how many are covered in Chapter 3. Second, urban deer density recommendations are based on the biological and cultural carrying capacities of each community. Both concepts are theoretical constructs without analytical support as covered in Chapter 3.

2.2.5 Deer Population Estimates: Money Not Well Spent

Attempting to estimate deer populations before defining goals is a critical error in community deer management. Many communities get preoccupied with knowing the number of deer in the area instead of focusing on the issues that are causing them to consider deer management in the first place. The question becomes, “How many deer should there be?” The fact is, deer problems are not defined by the number of deer, but should be defined by the impact they have on communities and the values of the residents they affect. Communities need to manage deer because residents agree that the number of deer-vehicle collisions, damage to home gardens and parks, or backyards that look like barnyards are unacceptable, not because there is a certain number of deer per square mile.

Deer population estimates are overrated as management objectives, are difficult to obtain, and usually confuse and divide communities. Attention should be focused on identifying and measuring the impacts of deer, not numbers of deer *per se*. Trends are important relative to your goals – if your goal is to reduce deer-vehicle collisions, you have achieved this goal when there are fewer accidents, not when there is an arbitrary density of deer in your community. The number of deer may be linked to deer-vehicle collisions, but you don’t need to know one to achieve the other.

Once deer management goals are established and agreed to by the community and/or its chosen governing body, population numbers may be helpful in determining the cost associated with effective deer removal strategies. In other words, it is helpful to get an idea of how many deer are in the area so that costs can be determined based on the number of animals that should be removed to have an effect. Given that a complete count of deer is unlikely, a simple, minimum count of deer in an area would be adequate to get started.

For example, once the goals are set and an agreement has been reached that a deer problem exists, a count is done and it is decided that 50 deer need to be removed. If the costs to remove just one deer are known, then a community can determine the total cost for this management action.

Remember, deer population numbers may be helpful in determining costs, but they are poor justification for taking management actions.

Even if deer numbers are ecologically balanced in your community, residents may still be experiencing unacceptable levels of conflict which will require some management action. The first step in community deer management is identifying goals and agreeing there is a problem that warrants management action. To be successful, communities should not justify deer management actions based on population estimates alone (*A Guide to Community Deer Management in Pennsylvania*. Version 2.4).

2.3 NATIONAL AND STATEWIDE ESTIMATES OF DEER NUMBERS

One of the most difficult tasks in urban deer management is to obtain credible statewide estimates of deer numbers. Estimated populations may be presented as ranges to denote the difficulty and levels of uncertainty in gathering an estimate over a large spatial scale. The above census techniques for urban communities cannot be applied directly on a statewide level of analysis. For the most part, population trends are based on harvest data (self-reported or check stations), aerial surveys, and hunter report cards. The statewide estimates for deer were obtained from many resources cited in Tables 2.1 through 2.4. These tables represent the first attempt to provide a national comparative overview of the status of information on cervid abundance in each state. It would be a fair assessment that no one knows exactly how many white-tailed, mule, and Key deer, or moose and elk reside in each state. There is no statistical confidence in the estimates. Factors that contributed to cervid population increases or decreases included climate changes, inclement weather (e.g., flooding, drought, hurricanes), abundance or loss of mast crops (e.g., acorns), habitat change, predators, parasites, and disease epidemics. Different sources of information usually produced variations by the thousands in estimates and conflicting reports on population trends in the same state. Some state estimates of particular cervids were dated by a decade or more with no indication of contemporary estimates forthcoming. This was usually the case for states that had small cervid populations, e.g., 20–40 moose in Wisconsin in 2003. For the most part, the estimates reported herein can only be used as cautionary relative comparisons. *This is not an exact science!* Yet, state DNR deer biologists are pressed professionally, publicly, and politically to provide a statewide cervid population estimate every year. "We believe political pressure and the size of elk herds are primary factors controlling why and how wildlife agencies are estimating populations" (Vermeire et al. 2004). Decades of trend data provide the statistical base to determine increasing, decreasing, stable, or uncertain evaluations of population trends.

TABLE 2.1
White-Tailed Deer Population Estimates (2014–2017) in the United States

States	Population Estimates[a]	Population Trends[b]
Alabama	1,600,000	+
Alaska	0	
Arizona	70–75,000	0
Arkansas	1,000,000	0
California	0	
Colorado	25,000	–
Connecticut	100,000	–
Delaware	45,000	–
Florida[c]	700,000	0
Georgia	1,270,000	–
Hawaii[d]	0	
Idaho	200,000	–
Illinois	650,000	–
Indiana	730,000	–
Iowa	400,000	0
Kansas	650,000	–
Kentucky	820,000	+
Louisiana	500,000	+
Maine	211,000	+
Maryland	214,000	–
Massachusetts	85–95,000	0
Michigan	1,750,000	+
Minnesota	1,000,000	+
Mississippi	1,750,000	0
Missouri	1,200,000	–
Montana	195,000	+
Nebraska	35–45,000	+
Nevada	0	
New Hampshire	85,000	–
New Jersey	101,000	+
New Mexico	10–15,000	–
New York	900,000	+
North Carolina	1,000,000	–
North Dakota	112,000	+
Ohio	660,000	0
Oklahoma	625,000	+
Oregon[e]	5,000	+
Pennsylvania	1,400,000	+
Rhode Island	16,000	+

(*Continued*)

TABLE 2.1 (CONTINUED)
White-Tailed Deer Population Estimates (2014–2017) in the United States

States	Population Estimates[a]	Population Trends[b]
South Carolina	725,000	0
South Dakota	550,000	+
Tennessee	950,000	0
Texas	4,000,000	+
Utah[f]	50–100	+
Vermont	157,000	+
Virginia,	810,000	–
Washington[g]	110,000–130,000	–
West Virginia	825,000	+
Wisconsin	1,200,000	+
Wyoming	70,000	+
Total	28.2–28.4 million	

[a] Deer Friendly. 2016. Available online at www.deerfriendly.com/home/about (Accessed June 2017) and (Vercauteren et al. 2011, Table 1).
[b] Trends: Increasing (+), decreasing (–), stable (0), unsure (?).
[c] Includes 600–700 Key deer (*O. v. clavium*).
[d] Exotic axis deer (*Axis axis*) is the most prevalent (10,000) cervid.
[e] Threatened Columbian white-tailed deer. USFWS Oregon F & W Office. Columbian white-tailed deer. Available online at www.fws.gov/oregonfwo/articles.cfm?id=149489413 (Accessed June 2017).
[f] Has 50–100 invasive aggressive white-tailed deer that appear to be a threat to mule deer
[g] Endangered Columbian white-tailed deer.

It would be a fair and testable assumption that cervid population size in each state will predict the probability of human/deer encounters in urban communities. Relevant data to test this assumption would include cervid numbers and the number of urban communities in each state reporting the need for an urban deer management program. The unit of observation would be by state. It seems intuitive that cervid distribution by state and numbers will correlate with the need for urban deer management plans. This assumption is supported by the inevitability that urbanization will result in human/cervid encounters.

White-tailed deer (N = 30.8 million) are the most abundant cervids in the Eastern United States. The only states lacking viable populations are

TABLE 2.2
Mule Deer Population Estimates in the United States in 2016

State	Population Estimates	Population Trends[a]
Alaska	333,000–346,000	+
Arizona	90,000–100,000	+
California	532,621	–
Colorado	424,000	–
Hawaii[b]	1,000–1,200	?
Idaho	120,000–200,000	+
Kansas	51,000	–
Montana	297,231	–
Nebraska	77,000	+
Nevada	94,000	–
New Mexico	80,000–100,000	–
North Dakota	14,000	+
Oklahoma	1,500–2,000	+
Oregon[c]	525,000–550,000	–
South Dakota	80,000–150,000	–
Texas	251,382	+
Utah	384,650	+
Washington	90,000–110,000	0 to –
Wyoming	353,000	–
Total	3.8–4.4 million	

[a] Trends: Increasing (+), decreasing (–), stable (0), unsure (?).
[b] Introduction of the Oregon black-tailed deer to the west of Kauai Island in 1961.
[c] Includes 300,000–320,000 black-tailed and 225,000–235,000 mule deer.

References

1. Mule Deer Working Group. Western Association of Fish and Wildlife Agencies. 2016. Range-wide status of black-tailed and mule deer. www.wafwa.org/Documents%20and%20Settings/37/Site%20Documents/Working%20Groups/Mule%20Deer/Publications2/2016_Mule_Deer_and_BTD_Status_Update_Final.pdf
2. Deer Friendly. 2016. Available online at www.deerfriendly.com/3home/about

Alaska, California, Hawaii, Nevada, and Utah (Figure 2.10, Table 2.1). The white-tailed deer estimate in Texas is at least 4.0 million, but ten other states, namely Alabama, Arkansas, Georgia, Michigan, Minnesota, Mississippi, Missouri, North Carolina, Pennsylvania, and Wisconsin have an estimated white-tailed deer population of 1 million plus.

The white-tailed deer distribution in the United States covers every ecoregion (Figure 2.11). Except for historical elk distributions, no other cervid demonstrates the level of adaptation required to survive in vastly different ecological regions. This comparison further attests to the inevitable

TABLE 2.3
Elk Population Estimates (2008–2017) in the United States

State[a]	Population Estimates	Population Trends[b]
Alaska[c]	900	0
Arizona	30,000–35,000	+
Arkansas	500	+
California	3,800	+
Colorado	278,000	+
Idaho	107,000	+
Illinois	23	0
Kentucky	10,000	0
Michigan	668	0
Minnesota	131	–
Missouri	125	+
Montana	149,000	+
Nebraska	2,600	+
Nevada	17,500	+
New Mexico	70,000	+
North Carolina	148	0
North Dakota	750–900	+
Oklahoma	800	+
Oregon	125,000	+
Pennsylvania	800	+
South Dakota (Black Hills)	7,200	+
Tennessee	300	+
Texas	1,600	+
Utah	79,000	+
Virginia	150–200	+
Washington	62,150	+
West Virginia	125	0
Wisconsin	200	+
Wyoming	93,000	+
Total	1–1.1 million	

[a] Missouri (2011), Tennessee (late 1990s), Virginia (2012), and West Virginia (2016) attempting elk (*C. e. nelsoni* not *C. e. canadensis* – now extinct) restoration programs with feasibility studies in Illinois, Maryland, and New York (Keller et al. 2015).

[b] Trends: Increasing (+), decreasing (–), stable (0), unsure (?) (http://newwest.net/topic/article/elk_populations_soaring/C41/L41/, Accessed June 2017).

[c] Eight herds on different islands (www.alaskannature.com/elk.htm, Accessed May 2017).

References

1. Billings Gazette. 2009. Wyoming's elk population 12 percent above goal. Available online at http://billingsgazette.com/news/state-and-regional/wyoming/wyoming-s-elk-population-percent-above-goal/article_e142f3a8-5779-59de-a7fd-93536e488895.html (Accessed January 18, 2017).

(*Continued*)

TABLE 2.3 (CONTINUED)
Elk Population Estimates (2008–2017) in the United States

2. Arkansas Game and Fish Commission. 2016. Elk in Arkansas. Available online at www.agfc.com/resources/Publications/elkinarkansas.pdf (Accessed May 2017).
3. Deer Friendly. 2016. Available online at www.deerfriendly.com/home/about
4. Fahrney, L. 2010. DNR says elk must "leave home." *Outdoor News*. Available online at http://outdoornews.com/2014/02/20/dnr-says-elk-must-leave-home/ (Accessed May 2017).
5. French, B. 2007. Growing elk population pits state against park. *Billings Gazette*. Available online at http://billingsgazette.com/news/features/outdoors/growing-elk-population-pits-state-against-park/article (Accessed May 2017).
6. California Department of Fish and Wildlife. Elk. Available online at www.wildlife.ca.gov/Conservation/Mammals/elk (Accessed January 18, 2017).
7. Gill, R.B., C. Gill, R. Peel, and J. Vasquez. 2013. Are elk native to Texas? Yes. Available online at http://circleranchtx.com/are-elk-native-to-texas-yes/ (Accessed March 24, 2017).
8. GoHunt Staff. 2014. Elk numbers across 6 states. Available online at www.gohunt.com/read/elk-numbers-across-6-states#gs.AjHE5eY (Accessed January 18, 2017).
9. Hunter, E.A. 2015. What states have elk? Available online at http://animals.mom.me/states-elk-4273.html (Accessed January 18, 2017).
10. Idaho Fish and Game. 2016. Elk population status by elk zone. Available online at https://idfg.idaho.gov/sites/default/files/seasons-rules-big-game-elk_2015-2016.pdf (Accessed January 18, 2017).
11. Michigan Department of Natural Resources. 2014. Michigan elk: past and present. Available online at www.michigan.gov/dnr/0,4570,7-350-79119_79147_81578-429560--,00.html.
12. Thomas, J.W. and D.E. Toweill. 1982. *Elk of North America – Ecology and Management*. Wildlife Management Institute. Stackpole Books, Harrisburg, PA. 698 pp.
13. U.S. Fish and Wildlife Service. 2014. Rocky mountain elk. Available online at www.fws.gov/refuge/Cold_Springs/Wildlife_Habitat/Elk.html (Accessed January 18, 2017).
14. Washington Hunter Ed. Course. 2015. Elk in Washington. Available online at www.hunter-ed.com/washington/studyGuide/Elk-in-Washington/20105001_700046864/ (Accessed January 18, 2017).
15. Wildlife Habitat Management Institute. 1999. American elk (*Cervus elaphus*) fish and wildlife habitat management. Leaflet No. 11. USDA. Natural Resources Conservation Service 8pp. Available online at www.nrcs.usda.gov. nrcs143_010000/ (Accessed April 2017).

occupation of white-tailed deer in urban areas, regardless of extreme variations in climate and habitat characteristics of different ecoregions. In addition, white-tailed deer have been transported from the United States to Jamaica, the British Isles, Czechoslovakia, Finland, Yugoslavia, New Zealand, Cuba, the Virgin Islands, Curacao, and other Caribbean Islands

TABLE 2.4
Moose Population Estimates (2015–2017) in the United States

	Population Estimates	Population Trends[a]
Alaska	175,000–200,000	+
Colorado	3,000	+
Connecticut	75–100	0
Idaho	10,000	–
Maine	60,000	+
Massachusetts	1,200	0
Michigan[b]	1,673	+
Minnesota	3,450	–
Montana	3,000–5,000	–
New Hampshire[b]	3,800	–
New York[c]	500–1,000	+
North Dakota	850	–
Oregon	70	+
Utah[b]	2,625	0
Vermont	2,200	–
Washington[c]	3,200	+
Wisconsin	20–40	0
Wyoming[b]	4,650	–
Total	292,300–318,000	

[a] Trends: Increasing (+), decreasing (–), stable (0), unsure (?). Estimated populations may be presented as ranges to denote the difficulty and levels of uncertainty in gathering an estimate over a large spatial scale.

[b] State listed as species of special concern. Does not include Isle Royale population.

[c] State protected species.

References

1. Alaska Department of Fish and Game. 2011. Moose (*Alces alces*) species profile. Available online at www.adfg.alaska.gov/index.cfm?adfg=moose.main
2. Associated Press. 2006. Moose population booming in parts of Washington, Idaho. Available online at http://missoulian.com/news/state-and-regional/moose-population-booming-in-parts-of-washington-idaho/article_98edf02f-73e0-55d9-ad72-03a8c959849f.html; https://idfg.idaho.gov/hunt/moose (Accessed January 17, 2017).
3. Associated Press. 2009. Wyoming moose numbers fall short. Available online at http://billingsgazette.com/news/state-and-regional/wyoming/article_eef704f1-9b7f-5812-b219-31b57a917a31.html (Accessed January 17, 2017).
4. Bannow, T. 2014. Oregon moose population higher than ever. *The Bulletin*. Available online at www.bendbulletin.com/home/1696456-151/oregon-moose-population-higher-than-ever
5. Beyer, D.E., S.R. Winterstein, P.E. Lederle, and J.J. Maskey. 2011. Moose in Michigan: history, biology, and considerations for hunting. Report to the Michigan Department of Natural Resources and Environment. Moose Hunting Advisory Council. Rpt. 350494.

(Continued)

TABLE 2.4 (CONTINUED)
Moose Population Estimates (2015–2017) in the United States

6. Deer Friendly. 2016. Available online at www.deerfriendly.com/home/about
7. Innes, R.J. 2010. *Alces americanus*. In: Fire Effects Information System. U.S. Department of Agriculture, Forest Service, Rocky Mountain Research Station, Fire Sciences Laboratory (Producer). Available online at www.fs.fed.us/database/feis/animals/mammal/alam/all.html/ (Accessed January 17, 2017).
8. Mass. Gov. Energy and Environmental Affairs. Moose. Available online at www.mass.gov/eea/agencies/dfg/dfw/fish-wildlife-plants/mammals/moose-in-mass-generic.html (Accessed January 17, 2017).
9. Matthews, P.E. 2012. History and status of moose in Oregon. *Alces* 48:63–66.
10. Timmerman, H.R. and A.R. Rodgers. 2017. The status and management of moose in North America – circa 2015. *Alces* 53:1–22.
11. Utah Division of Wildlife Resources. 2009. Utah moose statewide management plan. Available online at https://wildlife.utah.gov/hunting/biggame/pdf/moose_plan.pdf (Accessed May 2017).
12. Walsh, M. 2016. Sharp decline in Vermont moose herd raises questions about hunting. Available online at www.sevendaysvt.com/vermont/sharp-decline-in-vermont-moose-herd-raises-questions-about-hunting/Content?oid=3493802 (Accessed January 17, 2017).
13. Wattles, D.W. and S. DeStefano. 2011. Status and management of moose in northeastern United States. *Alces* 47:53–68.
14. Wyoming Game and Fish Department. Moose. Available online at https://wgfd.wyo.gov/WGFD/media/content/PDF/Habitat/SWAP/Mammals/Moose.pdf (Accessed May 2017).

(Chai 2003). "White-tailed deer should be defined by their ability to thrive over a mind-boggling variety of situations" (DeYoung 2011, p. 149). White-tailed deer are considered a "cosmopolitan" species because of their ability to live in such a diversity of habitats. There is some irony in realizing that humans also fit the definition of a cosmopolitan species, which makes it seem as if predestination predicted white-tailed deer/human cohabitation.

Urban deer appear to be their "own worst enemies." As thriving urban adapters, they are capable of temporary exponential population growth patterns leading to increased susceptibility and death due to environmental resistance factors. The primary environmental resistance factors affecting overabundant urban white-tailed deer are diseases (e.g., hemorrhagic), starvation, vehicle collisions, intentional herd reduction, and reduced fawn survival. Nevertheless, urban white-tailed deer have demonstrated the temporary impacts of these environmental resistance factors in preventing rebounds in population growth.

We tested the assumption that white-tailed deer numbers by state could be predicted by the square miles of urban areas in each state. This assumed

FIGURE 2.10 Geographic range of white-tailed (*Odocoileus virginianus*) ("White-tailed deer habitat." Whitetail Deer. Available online at https://whitetaildeerkellynjordan.weebly.com/habitat.html).

that the majority of deer in each state would be found in urban areas and that both variables, e.g., deer estimates (Table 2.1) and urban acres (U.S. Bureau of the Census 2010), are reasonably accurate. A Spearman rank correlation between the estimated white-tailed deer populations in each state, excluding Nevada and Utah, with square miles of urban and rural area was r = 0.6 and 0.2, p = 0.05, respectively. So there was some support for urban areas predicting state white-tailed deer numbers. The number of deer in each state was also correlated (r = 0.50, p = 0.05) with people per square mile. There may not be any relationship between correlation and cause, but apparently people and deer have a lot in common when it comes to the type of landscape(s) they prefer. Small state samples (<30) precluded our ability to test the same assumption for mule deer, moose, elk, and Key deer.

The endangered *Key deer* (N = 600–700) subspecies of white-tailed deer is concentrated (75%) on the Florida Keys Islands of Big Pine and No Name Keys (Figure 2.12). They became an inevitable urban occupant due to a surge of urbanization by humans from the mainland United States to the Florida Keys. Of note is how the Key deer have taken advantage of the alternative resources made available by the human invasive species (Figure 2.13). In September 2016, a screw worm epidemic infected the Key

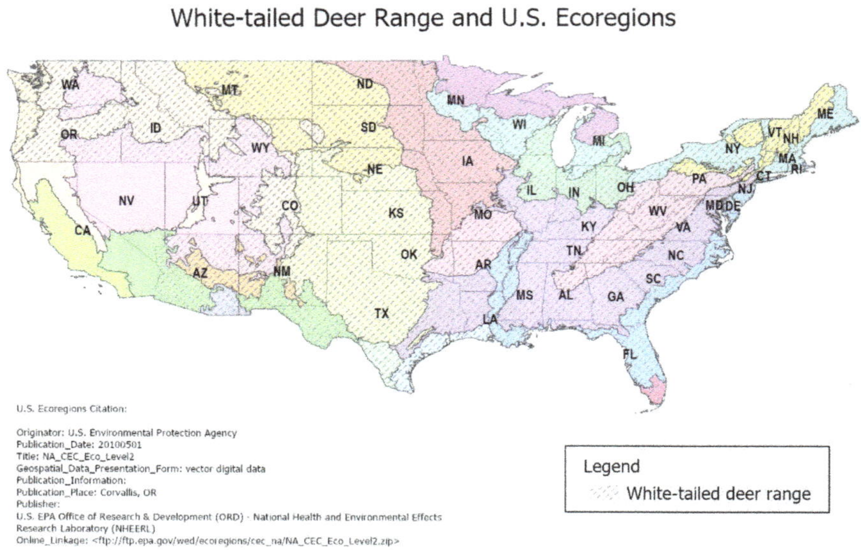

FIGURE 2.11 Note the extensive range of the white-tailed deer over nearly every ecoregion in the United States. Their adaptability has profound implications as a keystone species (courtesy Cassandra LaFleur Villarreal).

deer populations killing 135 before it was brought under control by April 2017. This epidemic demonstrated how susceptible endangered species are to the limiting factors of their environment.

Mule deer (N = 4.28 million) are concentrated in the western states and consist primarily of the Rocky Mountain subspecies (Table 2.2, Figure 2.14). However, major changes are occurring throughout much of mule deer range. The latest census indicated that the human population in the Western United States grew 16% from 2000 to 2010, which was 60% higher than the national average. The five states with the highest growth rates in the last decade (Nevada 35%, Arizona 25%, Utah 24%, Idaho 21%, and Texas 21%) were within the distribution of mule deer (U.S. Census Bureau 2010). This growth is placing increasing challenges on mule deer populations (Wakeling et al. 2015). The largest mule deer populations are in Oregon, California, Colorado, Utah, Wyoming, Alaska, and Montana. Nine states appear to be experiencing decreases in mule deer populations. Contributing factors may be the loss and change of habitat due to urban sprawl, predators, disease (e.g., chronic wasting disease (CWD)), highway and railroad kills, hunter overharvest (market hunting), fawn mortality, climate fluctuations (drought or extreme winter snow and cold), grazing competition with elk and domestic livestock, parasites (ticks), and nutritional losses. Of particular significance are impacts to habitat by human encroachment in the form of roads, fences and other barriers, subdivisions

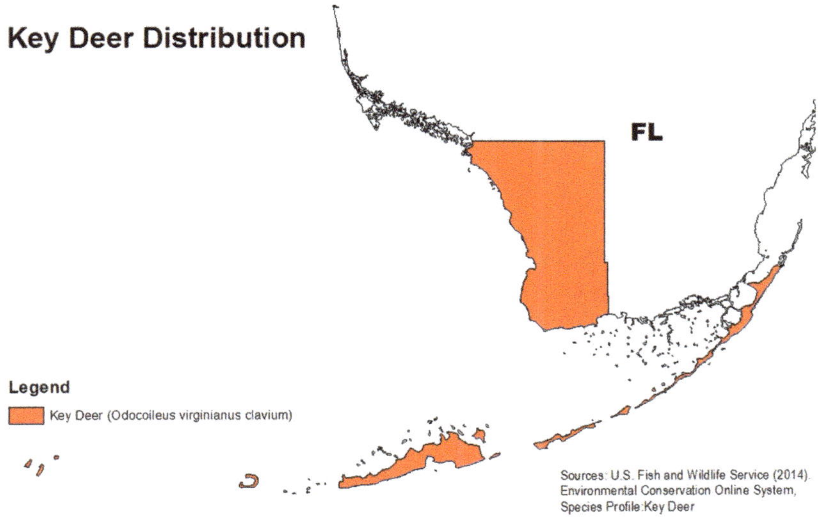

FIGURE 2.12 Geographic range of Key deer (*O. v. clavium*) (courtesy Cassandra LaFleur Villarreal; source: USFWS).

FIGURE 2.13 Key deer have adapted to urban existence by using alternative food sources provided by urban residents (courtesy Roel Lopez).

of land, and oil and gas drilling stand out ("Mule Deer Decline"). All of the previously listed causes are, at best, speculative and controversial, and probably operate in concert rather than as single factors. Also, credible studies that document causes and effects do not seem to exist (Gill 1999).

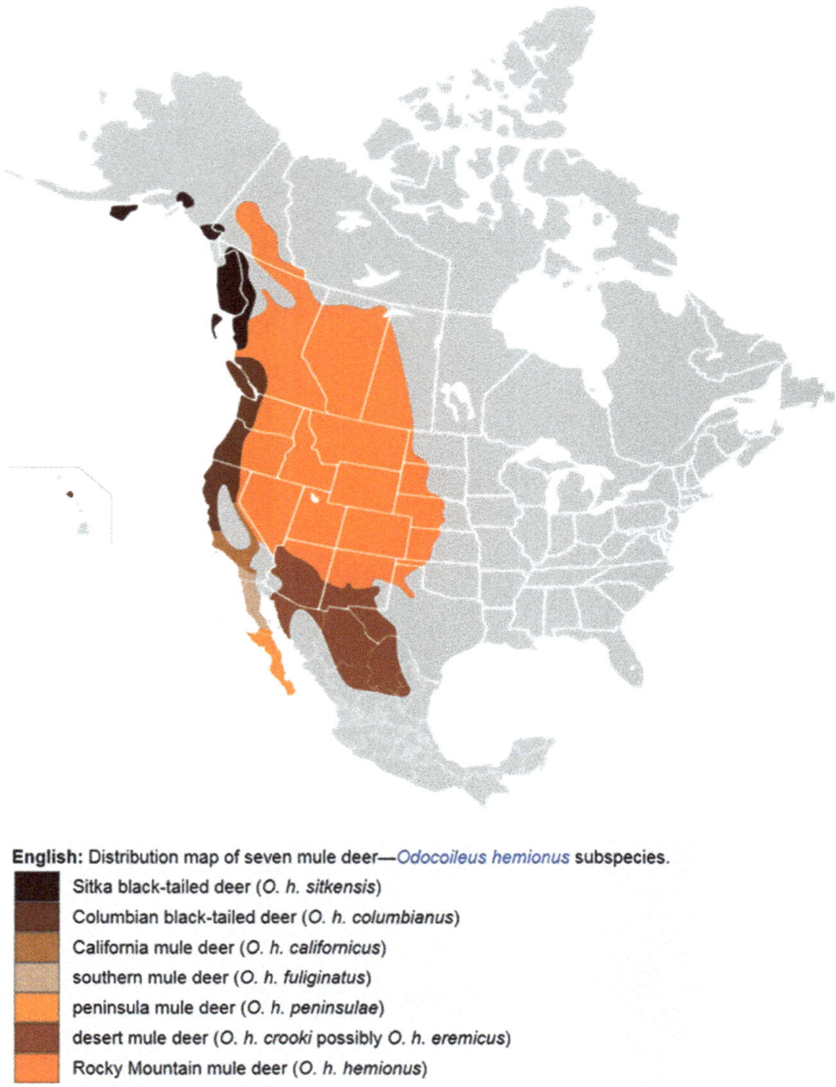

FIGURE 2.14 Geographic range of seven mule deer subspecies. (Wikipedia and www.gis.usu.edu/current_proj/muledeer.html)

Elk populations once exceeded 10 million and were distributed throughout the continental United States (Popp et al. 2014, citing Seton 1927). Today, six subspecies of elk number about 1 to 1.1 million (Table 2.3). The historical distribution of living and extinct elk subspecies demonstrated a near national coverage comparable to the white-tailed deer today (Figure 2.10). They were the most widespread ungulate in North America. Prior to European settlement there were huge herds of Eastern elk (*C. e. canadensis*) living in the dense hardwood forests that stretched from the Atlantic Coast to the Mississippi River. The human invasion of their natural habitat ultimately led to their total eradication through habitat conversion (forests to farmlands), over exploitation, purposeless slaughter of a relatively docile deer that was not overly fearful of humans, the usual human reaction to abundance of resources, and perhaps the inability of early American settlers to comprehend the long-term consequences of their actions. The Eastern elk was the first deer to be exterminated as a result of urban expansion. It should be noted that one of the trademarks of the westward expansion (1807–1910) was the extermination and/or depletion of wildlife. Other wildlife species that met the same fate as Eastern elk were the American buffalo (*Bison bison*) that lived east of the Mississippi River, whooping crane (*Grus americana*), passenger pigeon (*Ectopistes migratorius*), and Key deer. Key deer were already in a precarious situation in terms of population sustainability given their limited zoogeographic range in the Florida Keys. Comparatively, the range expansion and population explosion of white-tailed deer across America, because of urbanization, represent one of the most inexplicable contradictions in results.

The Roosevelt elk population, approximately 91,000 strong, ranges along the Pacific Coast from Northern California to Washington and Canada. The Rocky Mountain elk is the most plentiful of the four elk species (800,000–900,000 individuals) and occurs primarily in the mountain ranges east of the Cascade Mountains. However, because of its adaptive nature and variable diet, sustainable populations of Rocky Mountain elk have been successfully introduced or repatriated into historical habitats in many western and eastern states. Rocky Mountain elk now occur in parts of Alaska, Arizona, Arkansas, California, Colorado, Idaho, Kentucky, Michigan, Minnesota, Montana, Nevada, New Mexico, North Dakota, Oklahoma, Oregon, Pennsylvania, South Dakota, Texas, Utah, Washington, and Wyoming (Wildlife Habitat Management Institute 1999, Thomas et al. 1982). Elk (*C. e. nelsoni* not *C. e. canadensis* – now extinct) restoration programs have been attempted in Missouri (2011), Tennessee (late 1990s), Virginia (2012) and West Virginia (2016) with feasibility studies in Illinois, Maryland, and New York (Keller et al. 2015).

Moose (N = 292,000–318,000; Table 2.4) have expanded their range south to parts of the southern Rocky Mountains and spread into new areas of Washington, Oregon, Montana, Wyoming, Colorado, and Utah (Karns 2007, Table 2.5). Alaska is home to ten times the number of moose than in other states. Throughout their range, moose appear to be stricken by many factors limiting their ability to mature and produce offspring. For example, tick infestation, brain worm, carotid worm, liver flukes, climate change (cold to warmer), nutritional deficiencies, low calf survival, and predators are decreasing moose populations in some states. On the other

TABLE 2.5
Moose Subspecies and State Locations

Subspecies	State Locations
A. a. americana	Maine, New Hampshire, Vermont, Massachusetts, Rhode Island, Connecticut, and northern New York near the Adirondack Mountains. Population increasing.
A. a. andersoni	Michigan Upper Peninsula, northern Wisconsin, northern Minnesota, and northeastern North Dakota. Some state populations decreasing.
A. a. gigas	Alaska. The largest subspecies in North America (Franzman et al. 1978).
A. a. shirasi	Wyoming, Idaho, Utah, Colorado, Washington, Oregon, and Montana. Smallest subspecies in North America. Some state populations decreasing.

hand, healthy, well-fed cow moose will have twins as much as 75% of the time and triplets are common. The urban effects on moose populations are vehicle collisions and poaching. Urban moose are common in states with abundant moose populations, including Anchorage and Fairbanks, Alaska, and Bangor and Portland, Maine (see also Table 3.3 in the following chapter).

3 Overabundant Urban Deer

The problem of exploding deer populations has no attractive solutions.

George Johnson 2015

Managing an urban deer herd requires knowledge of the scope of the problem within urban communities. There are some similarities and differences as to how and why the deer management process is conducted in urban areas when compared to rural areas as discussed below. These comparisons address the most fundamental considerations in the deer management process.

It is common knowledge among urban residents and wildlife biologists working in public and private sectors that some species of deer, particularly white-tailed deer, are exceptionally abundant (impossible to provide real numbers) in urban communities. In order to provide some definition to relative numbers, we provided the first national synthesis of "locations" of urban deer herds that required some sort of management by some official. We had to use some unorthodox means of capturing the desired information (Tables 3.1 through 3.3) as discussed below in Section 3.2. The estimated number of white-tailed deer in each state (refer to Table 2.1 in the prior chapter) could be useful in suggesting positive correlations with urban deer locations within each state (Table 3.1). During this part of the study, we developed an overall perception that the scale of importance of this information to others was only the information specific to their own backyard, city, county, or state. Our "big picture" analysis, an original approach, with its national implications, should provide a more holistic examination of the scope of the urban deer problem given its extensive ramifications.

3.1 DEER MANAGEMENT IN URBAN AREAS WHEN COMPARED TO RURAL AREAS

Wildlife management in urban versus natural rural habitats is somewhat similar, but vastly different in many cases (Adams 2016, refer to Table 1.4 in a previous chapter). A close examination of these wildlife management comparisons provides an opportunity to realize the complexity of the process, and the implied academic and experiential training required of

TABLE 3.1
States, Counties, Cities, or Specific Locations Known To Have Urban White-Tailed/Mule/Key Deer Populations

States[a]	Cities[b]	Total
Alabama	Anniston, Auburn, Birmingham, Dauphin Island, Hoover, Huntsville, Mobile, Montgomery, Mountain Brook, Oak Mountain, Opelika, Oxford, Pell City, Tuscaloosa	14
Arizona[c]	Buckeye, Flagstaff, Kingman, Phoenix, Prescott, Tucson, Yuma	7
Arkansas	Bull Shoals, Cherokee Village, Diamondhead, Eureka Springs, Fort Smith, Heber Springs, Hope, Hot Springs Village, Lakeview, Little Rock, Magnolia, Russellville	12
California[c]	Alturas, Angel Island, Baldwin Park, Belvedere, Berkeley, Berryessa, El Cerrito, Los Angeles Metro, Lost Altos Hills, Marin County Cities, Monterey Peninsula Cities, Monterey, Napa County Cities, Nevada City, Oakland, Orange County Cities, Pacific Grove, Pasadena, Paso Robles, Ravenswood, Redding, Rocklin, Sacramento Suburban Areas, San Diego County Cities, San Diego, San Francisco Bay Area, San Jose, San Ramon Morgan Hill, Santa Rosa, Sonoma County, Villages (San Jose)	31
Colorado[c]	Alamosa, Boulder, Buena Vista, Canon City, Colorado Springs, Craig City, Denver, Durango, Elizabeth, Estes Park, Fort Collins, Loveland, Parker, Salida, The Pinery, Trinidad	15
Connecticut	Bethel, Bridgeport, Bridgewater, Brookfield, Danbury, Darian, Devil's Den Preserve, Easton, Greenwich, Groton Long Point, Mansfield Hollow, Monroe, Mumford Grove, New Canaan, New Fairfield, New Milford, Newtown, Norwalk, Norwich, Redding, Ridgefield, Shelton, Sherman, Stamford, Stratford, Trumbull, Weston, Westport, Wilton	30
Delaware	Brandywine, Burnett, Granite Shoals, Horseshoe Bay, Newark, Ocean View, Rose Valley, Wilmington	8
Florida	Bay County, Big Pine[d], Calhoun, Chiefland, Chipola River, Clewiston, Compass Lake Area, Dania, DeLand, Destin, Ft. Myers, Gadsden, Greenwich Audubon, Holmes, Jackson, Lake City, Lakeland, Liberty, Marianna, Mediterra, Monticello, No Name Key[d], Panama City, Sebring, Stat Rosa Beach, Tampa, Vero Beach, Walton, Washington	28

(Continued)

TABLE 3.1 (CONTINUED)
States, Counties, Cities, or Specific Locations Known To Have Urban White-Tailed/Mule/Key Deer Populations

States[a]	Cities[b]	Total
Georgia[g]	Athens, Atlanta, Bald Head Island, Barrow County, Big Canoe, Blythe Island, Brunswick, Cherokee County, Cobb County, Columbus, Dawson County, DeKalb County, Douglas County, Fayette County, Fayetteville, Forsyth County, Fulton County, Gainesville, Gwinnett County, Hall County, Henry County, Jekyll Island, Macon, Rockdale County, Roswell, Sandy Springs, Savannah, Valdosta, Walton County, Watkinsville	30
Idaho	Boise[c], Coeur d'Alene, Dalton Gardens, Helena, Pocatello, Sandpoint	6
Illinois	Carbondale, Channon Village, Chicago Metro Area, Cook County, Dundee Township, DuPage County, Edwardsville, Galena, Galesburg, Grundy County, Homer Glen, Joliet, Kane County, Lake County, Lake Forest, Macomb, Moline, Monticello, Mt. Vernon, Ogden Dunes, Peoria, Quincy, River Forest, River Woods, Rock Island, Rockford, Springfield, Urbana, Will County	29
Indiana	Beverly Shores, Bloomington, Boone County, Brookdale, Carmel, Chesterton, Clark County, Crown Point, Dune Acres, Evansville, Ft. Wayne, Gary, Hamilton County, Hidden Valley, Hobart, Indianapolis, Lafayette, LaGrange County, Lake County, Lawrenceburg, Marion County, Martinsville, Michigan City, Monroe County, Ogden Dunes, Porter County, Richmond, St. John, St. Joseph County, Valparaiso, Warsaw, Washington County	32
Iowa	Amana Colonies, Ames, Ankeny, Bettendorf, Cedar Rapids, Centerville, Clinton, Clive, Coralville/Johnson County, Council Bluffs, Davenport, Decorah, Des Moines, Dubuque, Fairfield, Galena, Green Valley State Park, Iowa City, Iowa Falls, Johnston, Keokuk, Marshalltown, Mason City, Moline, Muscatine, North Liberty, Oelwein, Oskaloosa, Ottumwa, Pleasant Hill, Polk County, Robins, Rock Island, Sioux City, Urbandale, Walnut Hill State Park, Waterloo	36
Kansas	Hutchinson, Kansas City Metro, Leavenworth, Louisville, Shawnee Mission Park, Shawnee, Topeka, Wichita	8

(*Continued*)

TABLE 3.1 (CONTINUED)
States, Counties, Cities, or Specific Locations Known To Have Urban White-Tailed/Mule/Key Deer Populations

States[a]	Cities[b]	Total
Kentucky	Anchorage, Ashland, Bowling Green, Covington, Elizabethtown, Fairdale, Fern Creek, Florence, Fort Mitchell, Fort Thomas, Henderson, Indian Hills, Lexington, Louisville, Morehead, Owensboro, Paducah, Pewee Valley, Prospect, Valley Station	21
Louisiana	Baton Rouge, Covington, Metairie, Monroe, Shreveport	5
Maine	Augusta, Bangor, Bar Harbor, Belfast, Biddeford, Cape Elizabeth, Cumberland, Falmouth, Farmington	9
Maryland	Abingdon, Accokeek at Gardner Park Road, Anne Arundel County, Antietam National Battlefield, Baltimore County, Bay Ridge, Bethesda, Betterton, Calvert County, Chevy Chase, College Park, Downs on the Severn, Fairfax County, Fort Washington, Fort Meade, Frederick County, Howard County, Lake Shore, Manassas National Battlefield, Monocacy National Battlefield, Montgomery Parks, Oregon Ridge Park, Pasadena Park, Prince George's County, Rock Creek Park, Rockville, Catoctin Mountain Park, Sligo Creek, Sunset Hills	28
Massachusetts	Andover, Barnstable, Blue Hills, Braintree, Burlington, Cambridge, Canton, Crane Beach in Ipswich, Dedham, Dover, Framingham, Greenwich, Hingham, Hopkins Memorial Forest in Williamstown, Lincoln, Lowell, Martha's Vineyard, Medfield, Medford, MetroWest, Milton, Naushon, Needham, Newton, Northampton Conservation Areas, Norwood, Plymouth, Sherborn, Shrewsbury, Springfield, Sudbury, Wayland, West Newbury, Westborough, Westwood, Weston, Westport, Worcester	36
Michigan	Alma, Ann Arbor, Barton Hills, Blandford Nature Center in Grand Rapids, Bloomfield Hills, Bloomfield, Broadview Heights, Chatham, Clarkston, Dearborn, Detroit Metro (Macomb, Oakland, and Wayne Counties), East Lansing, Ella Sharp Park and Cascades Golf Course in Jackson City, Farmington Hills, Fergus Fall, Genesee Township, Grand Forks, Grand Haven, Grosse Ile Township, Harbor Springs, Hillsdale City, Ironwood, Manistee, Meridian Township, Metamora, Midland, Milford, Mt. Pleasant, North Ottawa Dunes Park, Norton Shores, Ottawa County, Rochester Hills, Roscommon County, Royal Oak, Selfridge Air Force Base, Two Harbors	36

(Continued)

TABLE 3.1 (CONTINUED)
States, Counties, Cities, or Specific Locations Known To Have Urban White-Tailed/Mule/Key Deer Populations

States[a]	Cities[b]	Total
Minnesota	Anoka, Bemidji, Bloomington, Burnsville, Cook, Duluth, Eau Claire, Eden Prairie, Faribault, Fergus Falls, Forest Lake, Grand Haven, Grand Rapids, Hallock, Hermantown, Inver Grove, Lake Elmo, Lino Lakes, Little Canada, Mankato, Maplewood, Mendota Heights, New Port, New Ulm, North Oaks, Northland, Oakdale, Owatonna, Ramsey County, Red Wing, Redwood Falls, Roseville, Shoreview, Shorewood, St. Cloud, St. Croix State Park, St. John's Abbey Arboretum, St. Peter, Superior, Twin Cities Metro (Minneapolis and St. Paul), Two Harbors, Univ. of Minnesota Landscape Arboretum, West Lakeland, White Bear	44
Mississippi	Canton, Clinton, Jackson, Madison, Onalaska, Oxford, Starkville	7
Missouri	Ballwin, Blue Springs, Booneville, Branson, Bridgeton, Cape City, Cape Girardeau, Chesterfield, Chillicothe, Clarkson Valley, Columbia, Creve Coeur, Des Peres, Dueling, Ellisville, Eureka, Fulton, Hannibal City, Hardin Springs, Hazelwood, Hidden Valley Park, Hodge Park, Independence, Jefferson City, Kansas City Metro, Kirksville, Ladue, Lake Ozark, Lees' Summit, Little Blue Valley Park, Manchester, Marilyn Heights, Moberly, Osage Beach, Parkville, Powder Valley, River Front Park, Rolla City, Six Flags, Springfield, St. Louis suburban areas, Sunset Hills, Swope Park, Tiffany Springs Park, Town and Country, Village of Four Seasons, Wildwood	47
Montana[c]	Bitterroot Valley, Bozeman, Butte, Colstrip, Flathead Valley, Fort Benton, Gardiner, Glasgow, Glendive, Great Falls, Helena, Libby, Missoula, Roundup, Troy, Whitefish	16
Nebraska[c]	Bellevue, Eugene T. Mahoney, Fontenelle Forest, Lincoln, Omaha, Platt River State Park, Schramm State Parks	7
Nevada[c]	Carson City, Eureka (Carson City), Galena, Garnerville, Genoa, Kings Canyon (Carson City), Lamoille, Minden, Pioche, Reno, Spring Creek	11
New Hampshire	Allenstown, Atkinson, Balch Hill Area (Hanover), Bernardsville, Boscawen, Bow, Canterbury, Chichester, Concord, Dunbarton, Epsom, Flat Rock Brook Nature Center, Franklin, Freehold Township, Hanover, Henniker, Hopkinton, Jersey City, Little Falls, Loudon, Moultonborough, Pembroke, Pittsfield, Saddle River, Salisbury, Trenton, Warner, Weare, Webster	29

(*Continued*)

TABLE 3.1 (CONTINUED)
States, Counties, Cities, or Specific Locations Known To Have Urban White-Tailed/Mule/Key Deer Populations

States[a]	Cities[b]	Total
New Jersey[h]	Allen Dale, Ash Brook Reservation, Bayonne, Bedminster, Berkeley Heights, Bernard, Bernards Township, Bernardsville, Byram, Cheesequake, Clinton Township, Colts Neck, Cranford, Denville, Dover, Eagle Rock, Englewood, Essex County, Fairfax County, Flat Rock Brook Nature Center, Franklin Lakes, Gero Park (Millburn), Hanover, Harding Township, Haworth, High Bridge, Hillsborough, Hillsdale, Hopatcong, Hopewell Township, Howell, Lenape Park, Lewis Morris Park, Linden, Livingston, Loantaka, Lorrimer Sanctuary, Lower Township, Madison, Maplewood, Medford, Mendham, Mercer Meadows Park, Milburn, Monmouth County, Morris County, Morris, Morristown Park, Mount Olive, Mountainside, New Providence, Newark, Nomahegan Park, North Bergan, North Caldwell, Ocean View, Old Bridge Park, Princeton, Randolph, Raritan, Readington, Reeves-Reed Arboretum (Summit), Roackaway Township, Rutherford, Saddle River, Scotch Plains, Shrewsbury, South Orange, Springfield, Tenafly Nature Center, Union County, Verona-Cedar Grove, Warren Township, Washington Township, Watchung, West Armwell, West Caldwell, West Milford, West Orange	79
New Mexico[c]	Albuquerque, Farmington, Raton, Roswell, Ruidoso, Santa Fe, Silver City	7
New York	Albany, Amherst, Athens, Avon, Batavia, Bethlehem, Binghampton, Bridgehampton, Bronx Park, Brookhaven National Lab, Buffalo, Busti, Cayuga Heights, Celoron, Cheektowaga, Clarence, Cornell University, Cutchoque, Delmar, Dobbs, East Hampton, East Marion, East Side, Eastern Long Island, Ellicott, Elmira, Erie County, Erwin, Fairfax County, Fayetteville, Ferry, Fire Island, Genesee, Hamilton, Hanover, Hastings-on-Hudson, Irondequoit, Ithaca, Jamestown City, Lackawanna, Lansing, Lloyd Harbor, Long Island, Mamaroneck, Manhattan, Manlius, Mountain Lakes Park in North Salem, Mt. Sinai, Muscoot Farm and Lasdon Park and Arboretum in Somers, New York City Parks, New York metro, Newark, Niagara Falls, North Fork, North Haven, North Tonawanda, Old Field, Olde Towne, Pelham Bay Park, Poquott, Port Jefferson, Port Townsend, Poughkeepsie, Pound Ridge, Reservation in	93

(*Continued*)

TABLE 3.1 (CONTINUED)
States, Counties, Cities, or Specific Locations Known To Have Urban White-Tailed/Mule/Key Deer Populations

States[a]	Cities[b]	Total
	Cross River, Riverdale Park, Riverhead, Rochester Institute of Technology, Rochester, Rye, Shelter Island, South Hampton, Southland, Southold, Springville, Staten Island, Suffolk County, Syracuse, Tarrytown, Teatown, Tompkins, Town of Riverhead, Town of Varik, Trumansburg in the Finger Lakes, Tuxedo Park, Van Corlandt Park, Vassar, Ward Pound Ridge, West Elmira, West Seneca, Westchester County, White Plains, Willamsville Cemetery	
North Carolina	Albemarle, Bald Head Island, Belmont, Boxborough, Burgaw, Carrboro, Chapel Hill, Charlotte, Clayton, Duke Forest, Durham, Emerald Isle, Fayetteville, Gastonia, Greensboro, Iredell, Jacksonville, Jonesville, Landis, Lejeune, Matthews, Mecklenburg, North Wilkesboro, Olde Point, Pinehurst, Raleigh, Sanford, Smithfield, Southern Pines, Stoneville, Wake Forest	31
North Dakota[c]	Bismarck, Fargo, Fort Ransom, Grand Forks, Jamestown, Minot, Stetter, White Hall	8
Ohio[e,i]	Akron, Amberley Village, Athens, Avon Lake, Bay Village, Beachwood, Bedford, Berlin Heights, Brecksville, Broadview Heights, Bryan, Bucyrus, Charlies Island, Chillicothe, Cincinnati, Clarington, Cleveland Heights, Cleveland Metroparks, Clifton, Columbus, Cuyahoga Valley National Park, Dayton, Dublin, Eastlake, Elyria, Fort Wright, Gahanna, Gambier, Granville, Highland Heights, Hillsboro, Hinckley, Holden Arboretum, Hudson City, Hunting Valley, Indian Hill, Lake Metroparks, Lancaster, Lyndhurst, Madison Township, Mayfield Heights, Medina, Mentor, Meridian, Metroparks by City, Mt. Vernon, New Albany, Newark, North Olmsted, North Ridgeville, North Royalton, Oakwood, Oelwein, Olmsted Falls, Ottawa Hills, Oxford, Parma Heights, Parma, Pepper Pike, Powell, Presque Isle, Seven Hills, Shaker Heights, Sheffield Village, Silver Lake, Solon, South Euclid, Strongsville, Summit Metroparks, Swanton Township, Toledo Metro Parks, West Lake, Westerville, Willoughby Hills, Wilmington, Wyoming, Youngstown, Zanesville	78
Oklahoma	Edmond, Grove, Norman, Oklahoma City, Shawnee, Stillwater, Tulsa	7

(Continued)

TABLE 3.1 (CONTINUED)
States, Counties, Cities, or Specific Locations Known To Have Urban White-Tailed/Mule/Key Deer Populations

States[a]	Cities[b]	Total
Oregon[c]	Ashland, Baker City, Bend, Burns, Caldera Springs, City of Sisters, Corvallis, Eagle Point, Enterprise, Eugene, Grants Pass, Jacksonville, John Day, Joseph, Klamath Falls, La Grande, Lake View, Lincoln City, Lostine, Madras, Marietta, Medford, Portland, Prineville, Redmond, Roseburg, Union City, Wallowa	28
Pennsylvania	Abington Lorimar Park, Altoona, Bedminster, Berks County, Bethel Park, Bethleham, Borough of Fox Chapel, Bowman's Hill Wildlife Preserve, Chambersburg, Cherry Valley, East Goshen Township, Karebrook Nature Study Area, Mt. Lebanon, Newton, Norristown Farm Park, Peters Township, Fairmount Park, Philadelphia, Pittsburg, South Hills, Township of Pine, Valley Forge Park	22
Rhode Island[f]	Barrington, Block Island, Bristol, Burrillville, Central Falls, Charlestown, Coventry, Cranston, Cumberland, East Greenwich, East Providence, Exeter, Foster, Glocester, Hopkinton, Jamestown, Johnston, Lincoln, Little Compton, Middletown, Narragansett, Newport, New Shoreham, North Kingston, North Providence, North Smithfield, Pawtucket, Portsmouth, Richmond, Scituate, Smithfield, South Kingston, Tiverton, Warren, Warwick, West Greenwich, West Warwick, Westerly, Woonsocket	40
South Carolina	Beaufort County, Belfair, Berkeley Hall, Callawassie Island, Charleston, Colleton River Plantation, Fairview, Fripp Island, Haig Point, Hilton Head Island, Huntersville, Indian Trail, Indigo Run, Long Cove Club, Mineral Springs, Mint Hill, Moss Creek, Palmetto Bluff, Palmetto Hall, Pineville, Sea Pines Plantation, Seabrook Island, Stallings, Tega Cay, Unionville, Waxhaw, Weddington	27
South Dakota[c]	Aberdeen, Custer, Edgemont, Ellsworth AFB, Hot Springs, Lead, Pierre City[c], Rapid City[c], Sioux Falls, Spearfish, Sturgis, Watertown, Whitewood	13
Tennessee[g]	Ashland City, Bristol, Chattanooga, Columbia, Dickson, Franklin, Germantown, Hendersonville, Jackson, Kahite, Knoxville, Lawrenceburg, Lebanon, Manchester, Memphis, Mt. Juliet, Murphysboro, Nashville, Oak Ridge, Oxford, Teatown, Tellico	22

(Continued)

TABLE 3.1 (CONTINUED)
States, Counties, Cities, or Specific Locations Known To Have Urban White-Tailed/Mule/Key Deer Populations

States[a]	Cities[b]	Total
Texas	Alpine[c], Arlington, Bandera, Barrington in Kingwood, Benbrook, Boerne, Bryan, Burnet, Canyon Lake, Carrollton, City of Grey Forest, Collin, Dallas Metro, Dallas, Delaware Springs, El Paso[c], Elm Creek Village (San Antonio), Fair Oaks Ranch, Fort Worth Metro, Fox Fire (College Station), Galveston, Garden Ridge, Granite Shoals, Grapevine, Harker Heights, Hayes, Hill Country Village, Holly Lake, Hollywood Park (San Antonio), Horseshoe Bay, Hudson Oaks, Johnson Space Center (Houston), Katy, Kerrville, La Bota Ranch (Laredo), Lake Tanglewood[c], Lakeway (Austin), Lakewood Hills (Austin), Laredo, Leander, Lewisville, Liberty Hill, Montgomery County, Montgomery Trace, Morgan's Point Resort, Navasota, New Braunfels, Northwest Hills, Port Mansfield, Ridge Lake Shores, Riverhill (Kerrville), Rockwall, Roseheart (San Antonio), Ruby Ranch, Salado, San Angelo, San Marcos, Shady Oaks, Spicewood, Sun City (Georgetown), Texas City, The Colony, Walden (Conroe), Weslaco, Westlake Hills, Woodlands (Houston)	66
Utah[c]	Alpine, Bountiful, Brigham City, Brigham Young University, Cache Valley, Cedar City, Centerville City, Draper, Farmington, Fruit Heights, Herriman City, Highland, Holladay, Hyde Park, Kaysville, Layton, Logan, Mapleton, Moab, Monticello, Newton, North Logan, Ogden, Orem, Pleasant View, Providence, Provo, River Heights, Roosevelt, Salt Lake City, Sandy City, Smithfield City, South Weber, Springville	34
Vermont	Bennington, Burlington, Essex Junction, Shelburne	4
Virginia[g]	Altavista, Amherst, Antietam National Battlefield, Alexandria, Arlington, Ashburn Park, Banshee Reeks Nature Preserve, Bedford City, Blacksburg, Blackstone, Boones Mill, Bristol, Charlottesville, Chatham, Chesapeake, Chesterfield County, Christiansburg, City of Richmond, Clifton, College of William and Mary, Colonial Heights, Covington, Chesapeake, Culpeper, Danville, Dunlora, Emporia, Fairfax County, Fairfax, Falls Church, Farmville, Franklin, Fredericksburg, Front Royal, Galax, Hampton,	85

(Continued)

TABLE 3.1 (CONTINUED)
States, Counties, Cities, or Specific Locations Known To Have Urban White-Tailed/Mule/Key Deer Populations

States[a]	Cities[b]	Total
	Harrisonburg, Hopewell, Hurt, Independence, Irvington, James City County, Kenbridge, Lake Ridge, Leesburg, Lexington, Loudon County, Lynchburg, Manassas, Mantua, Marion, Martinsville, Montgomery County, Mt. Vernon, Newport News, Norfolk, Pearisburg, Poquoson, Prince William County, Radford, Reston, Richlands, Richmond, Roanoke County, Roanoke, Rocky Mount, Saltville, Smithfield, Smith Mountain Lake, Staunton, Stuart, Suffolk, Town of Bedford, Town of Halifax, Town of Pulaski, Town of Tazewell, Victoria, Virginia Beach, Waynesboro, Williamsburg, West Point, Winchester, Wytheville, York Town, York County	
Washington[c]	Bellingham, Clinton, Coupeville, Colville City, Island County, Lakewood City, Oak Harbor, Olympia, Ponderosa (Spokane Valley), Port Townsend, Seattle, Spokane, Whidbey Island, Yakima	14
West Virginia	Beckley, Buckhannon, Charleston, Clarksburg/Bridgeport, Elkins, Elkview, Fairmont, Harper's Ferry, Huntington, Kanawha City, Marlinton, Martinsburg/Shepherdstown, Martinsburg, Morgantown, Mountwood, Parkersburg, Ripley, Ronceverte, Shepherdstown, South Charleston, Twin Falls, Weirton, Wheeling	23
Wisconsin[h]	Allouez, Alma, Ashland, Ashwaubenon, Bayport, Biron, Black River Falls, Blaine, Brookfield, Buchanan, Cedar Rapids, Cedarburg, Darboy, Eau Claire, De Pere, Delafield, Door County, Eau Claire, Eldorado, Elm Grove, Fitchburg, Fond du Lac, Fox Point/Bayside, Germantown, Grand Rapids, Green Bay, Holland, Hudson City, Kohler, La Crosse, Lake Delton, Madison, Marinette, Marshfield, Melrose, Menomonee Falls, Mequon, Merrill, Middleton, Milwaukee, Neillsville, New Port, New Richmond, Onalaska, Oshkosh, Owen, Pewaukee City, Portage City, Reedsburg City, Rhinelander, River Hills, River Valley, Sheboygan, Shell Lake, Stevens Point, Sturgeon Bay, Superior, Thiensville, Washburn, Waupaca, Wisconsin Rapids	61

(*Continued*)

TABLE 3.1 (CONTINUED)
States, Counties, Cities, or Specific Locations Known To Have Urban White-Tailed/Mule/Key Deer Populations

States[a]	Cities[b]	Total
Wyoming[c]	Buffalo, Casper, Cody, Craig, Diamondville, Green River, Jackson, Lander, Mountain View, Rawlins, Sheridan, Story, Sundance, Thermopolis	14
Total Locations		1,309

Notes:

[a] The above list of locations probably vastly underestimates the universe of locations. No state DNR keeps a master list of urban locations that require some form of urban deer management.

[b] White-tailed deer considered abundant in suburban/urban areas.

[c] Probably mule deer only or includes mule deer in some locations.

[d] Home of urban Key Deer

[e] All cities, towns, and other populated areas (N = 220) in Cuyahoga County. "Nearly every community in Cuyahoga County is debating how to handle deer" (Kilpatrick, M. 2015. "7 Cuyahoga County cities unite to ask voters to curb deer population." Cleveland.com). At least 54.2% of Cuyahoga County has some form of lethal control. All cities in Ohio have urban deer (Conover 1995).

[f] White-tailed deer are widespread in Rhode Island and were harvested in 31 of 39 towns and cities (Tefft, B.C. 2016. Rhode Island White-Tailed Deer Status Report. Rhode Island Department of Environmental Management. Division of Fish and Wildlife. West Kingston, RI, pp. 16, table 9).

[g] Respondents reported "statewide" or "nearly every city."

[h] The DNR fired the first salvo in Wisconsin's urban deer war with the establishment of metro deer management units in 1992. That year, units 59M (La Crosse), 76M (Madison) and 77M (metro Milwaukee) were created. Units 60M (Hudson) and 64M (Green Bay) followed in 1996, and Unit 1M (Superior) in 2001. In 2009, deer removal permits were issued to 32 municipalities, 20 airports, and 2 nature centers across Wisconsin (Small, D. 2010. Wisconsin's Deer Dilemma. September Game and Fish Magazine.)

[i] In 1992 Ohio DNR created five Urban Deer Hunting units including the Columbus, Cincinnati-Dayton, Cleveland-Akron, Youngstown, and Toledo Urban Deer Units (Bleech, M. 2010. "Ohio's Urban Deer Units – Are They Working." October, *Game and Fish* Magazine).

References

1. www.deerfriendly.com/deer/
2. Conover, M.R. 1995. What is the urban deer problem and where did it come from? pp. 11–18. *In* J.B. McAninch, ed., Urban deer – a manageable resource? Proc. 1993 Symp. North Central Section, The Wildlife Society. pp. 175.
3. Adams, C.E. 2004. National assessment of urban deer (unpublished data).

TABLE 3.2
States, Counties, Cities, or Specific Locations Known To Have Urban Elk Populations

States[a]	Cities[b]	Total
Arizona	Flagstaff, Payson, Star Valley	3
Arkansas	Buffalo National River, Hasty, Ponca	3
Colorado	Bobcat Ridge Natural Area (Fort Collins), Craig, Estes Park, Woodland Park	4
Idaho	Boise, Garden Valley, Lewiston	3
Kansas	Fort Riley Army Base, Johnson, Junction City, Kingman County, Pratt County, Reno County	6
Kentucky	Harlan, Hazard, Inez, Pikeville, Pippa Passes, Prestonsburg, Stoney Fork	7
Michigan	Atlanta, Gaylord City Elk Park	2
Missouri	Lone Elk Park Animal Sanctuary (St. Louis County), Peck Ranch Conservation Area (Shannon and Carter Counties)	2
Montana	Dillon, Kalispell, Missoula, White Sulphur Springs	4
Nevada	Ely	1
New Mexico	Cloudcroft, Ruidoso	2
Oregon	Astoria, Eugene, Gearhart, Lincoln City, Manzanita, Portland, Sea Side	7
Pennsylvania	Benezette, Elk County, Grant, Sinnemahoning Stated Park, Weedville	5
Tennessee	Crossville, Huntsville, Oak Ridge	3
Texas	Waco (2015 single sighting), Wheeler County	2
Utah	Park City	1
Total Locations		55

Notes:

[a] Small state elk populations precluded observations in urban areas.

[b] Elk often sighted in adjacent rural areas or park areas.

persons involved in the wildlife management process is necessary. The bottom line is that urban wildlife management involves so much more than just "animal damage control."

3.1.1 Similarities

Similarities in the wildlife management process when conducted in urban or rural habitats:

1. Involve game, nongame, exotic, and threatened/endangered species.

TABLE 3.3
States, Counties, Cities, or Specific Locations Known To Have Urban Moose Populations

States[a]	Cities[b]	Total
Alaska	*Found in much of Alaska.* Anchorage, Eagle River, Fairbanks, Homer, Juneau, Seward, Wasilla	7
Colorado	Arvada, Aspen, Boulder, Breckenridge, Broomfield, Colorado Springs, Denver, Fort Collins, Grand Lake, Lakewood, Steamboat Springs	11
Connecticut	*Moose sightings in urban communities are rare.* Barkhamsted, Bloomfield, Broomfield, Colebrook, Eastford, Enfield Winsted, Goshen, Granby, Hartland, Mansfield, New Britain, New Hartford, New Milford, Plainville, Ridgefield, Union, Winchester	17
Idaho	Boise, Coeur d'Alene, Hailey, Idaho Falls, Ketchum, Moscow, Potlatch, Sandpoint, Wallace	9
Massachusetts	Boston, Ludlow, Pelham, Quabbin Reservoir, Royalston, Worcester	6
Maine	Ashland, Bangor, Caribou, Fort Kent, Houlton, Medway, Millinocket, New Sweden, Portland, Presque Isle, Rangeley, Rumford, Woodland	13
Michigan[b]	Isle Royale National Park[c]	
Minnesota[b]	Duluth, Isabella	2
Nebraska[a]	Lincoln County, North Platte	2
New Hampshire	Colebrook, Concord, Conway, Dalton, Farmington, Keene, Manchester, Pittsburg	8
New York	Bellmont, Ellenburg, Franklin, Saranac, Troy, Westchester	6
North Dakota[b]	Bismarck, Fargo, Minot, Mohall, Williston	5
Oregon	Elgin, Weston	2
South Dakota	Alexandria, Avon, Brandon, Chester, Colton, Dell Rapids, Dimock, Elk Point, Madison, Parkston, Sioux Falls	11
Utah	Eden, Huntsville, Logan, North Ogden, Ogden, Park City, Provo, Rexburg, Richmond, Salt Lake City, South Weber, Springville	12
Vermont	Andover, Belvidere, Bloomfield, Brighton, Burlington, East Haven, Lemington, Lewis, Montpelier, Newark, Westford	11
Washington	Indian Trail, Liberty Lake, Palouse, Post Falls, Pullman, Spokane, Wenatchee, East Wenatchee, Royal City	9
Wisconsin[b]	Ashland Northland College, Burlington, Montpelier, Shawano County, Westford	5
Wyoming	Cody, Jackson Hole, Lander, Pinedale	4
Total Locations		140

Notes:

[a] Moose spotted in North Platte, NE, November 2015, and in northeastern Nevada, 2013. Most moose sightings occur when they just "wander into the neighborhood, stay awhile, leave, or are chased out of town."

(*Continued*)

TABLE 3.3 (CONTINUED)
States, Counties, Cities, or Specific Locations Known To Have Urban Moose Populations

[b] Rapid declines in moose populations due to climate change (too warm too long), internal and external parasites, disease, predation (wolf, bears, mountain lions), low calf survival, loss of natural habitat (old-growth spruce-fir forests), physiology (lack of sweat glands and thermoregulation during hot weather), moose road kills, and malnutrition.

[c] Moose population booming (N = 1,600) due to the loss of wolves. No urban connection though (see predator prey cycle – Figure 3.4)

References

1. Anderson, Jim. *Source Weekly*. April 2016. "Oregon's Moose Count."
2. Boughton, Kathryn. *The Litchfield County Times*. September 2013. "Moose on the Loose: Increased Presence in Connecticut."
3. Charles, Candese. KFYRTV. March 2017. "Moose on the loose in Mohall, N.D."
4. Cura, Jaimie. Mansfield Patch. September 2015. "Moose Sighting in Mansfield."
5. Hoffman, Justin D., Genoways, H.H., and Choate, J.R. January 2006. "Long-distance dispersal and population trends of moose in the central United States."*Alces*. 42. pp.115–131.
6. Kelly, J. Williston Herald. January 2017. "Moose."
7. Koshmrl, M. Jackson Hole News & Guide. March 2014. "Jackson Hole moose numbers looking up."
8. Levine, Justin A. Lake Placid News. January 2016. "DEC hopes to learn moose population numbers in the Park."
9. Lincoln Journal Star. November 2015. "Moose spotted in Nebraska; hunters urged to be cautious."
10. Minnesota Department of Natural Resources. Moose project archives.
11. Moritz, Katie. Juneau Empire. September 2013. "Over the years, moose sightings becoming more common."
12. NBC Connecticut. May 2015. "Moose Spotted in Bloomfield.
13. Peterson, E. Fox11. September 2015. "DNR: Moose spotted in Shawano Co. likely came from U.P."
14. Remington, Tom. May 2016. "Evidently Global Warming is Causing More Moose in Connecticut.
15. Shaw, Mitch. Standard Examiner. August 2017. "North Ogden looking to curb urban deer population, bows and arrows an option."
16. Stephen, Kaime. The North Platte Telegraph. December 2015. "Moose no longer on loose in N. Platte.
17. Valley News. October 2015. "Moose roaming north Fargo."
18. Vermont Fish & Wildlife. 2016. "2016 Vermont Wildlife Harvest Report-Moose."
19. Warden, Beth. November 2016. "Moose on the Loose in South Dakota."
20. Wattles, D.W. and DeStefanco, S. University of Massachusetts. February 2009. "Progress Report: Movement and Landscape Pattern Use of A Colonizing Moose Population in Massachusetts."

2. Use the standard wildlife management procedures (*Wildlife Techniques Manual – Management*, Silvy 2012).
3. Require input, participation, and oversight by state and/or federal agencies.
4. Require the assistance of wildlife biologists trained in appropriate wildlife management degree programs.
5. Rely on potential economic gains and losses as the primary catalysts for management actions.
6. Address proactive and reactive wildlife management goals.
7. Use professional (e.g., journal articles) and popular outlets (e.g., magazines and newspapers, TV, and internet sites) to disseminate the status of information to the *whole* community.

3.1.2 Differences

The differences in the wildlife management process in urban when compared to rural habitats are summarized below.

1. Urban, when compared to rural habitats, have less plant and animal diversity. This results in wildlife management programs that focus on a few species (urban exploiters) that are urban adaptable, or well-suited to an urban environment, e.g., deer.
2. The population sizes of some urban adaptable species are larger than the same species' populations in rural habitats.
3. There are more dedicated state and federal funding sources for wildlife management in rural when compared to urban habitats. This has led to limited academic and agency interest in urban wildlife management.
4. There is a large and well established focus for research, management, and education programs in rural habitats based on historical precedent. Comparatively, these programs in urban habitats are in the early stages of development.
5. Layers of jurisdiction (e.g., federal, state, county, municipal) increase with proximity to urban centers, but decrease as wildlife management activities become more rural-oriented.
6. Urban wildlife management is conducted within small scales of analysis, with many legal and physical impediments, and in highly fragmented landscapes. Rural wildlife management is conducted within large scales of analysis, with few legal and physical impediments, and in less fragmented landscapes.
7. Academic and practical training in the "people factors" is an imperative prerequisite in the conduct of urban wildlife management. This includes conflict resolution; awareness of public

attitudes, activities, knowledge, and expectations concerning the wildlife management process in their neighborhoods; public education; and identification and inclusion of all stakeholder groups (see Figure 3.1). These are daunting tasks but not significant considerations in rural wildlife management.

8. Wildlife managers have to deal with a more heterogeneous set of public attitudes and expectations related to wildlife in urban when compared to rural habitats (Figure 3.1).
9. Urban residents have high inclusion needs in the wildlife management process and demand participation (Figure 3.1). Rural residents are included in the process automatically by virtue of ownership of wildlife habitat and compatibility of management goals (e.g., recreational hunting) with state and federal game agencies.
10. Due to concentrations of people and wildlife in urban communities, there is a higher potential for a threat to public health from zoonotic diseases and parasites when compared to rural areas.
11. Usually the overarching management goal is to *reduce* or *sustain* overabundant wildlife populations in urban and rural areas.
12. There is a growing trend toward privatization and commercialization of wildlife management in urban areas, e.g., animal damage control (Table 4.4). The majority of wildlife management efforts in rural areas are coordinated through landowners and state and federal agencies.
13. It takes longer to complete management activities in urban areas, when compared to rural areas, which may task the patience of urban residents, and preclude successful management goals.

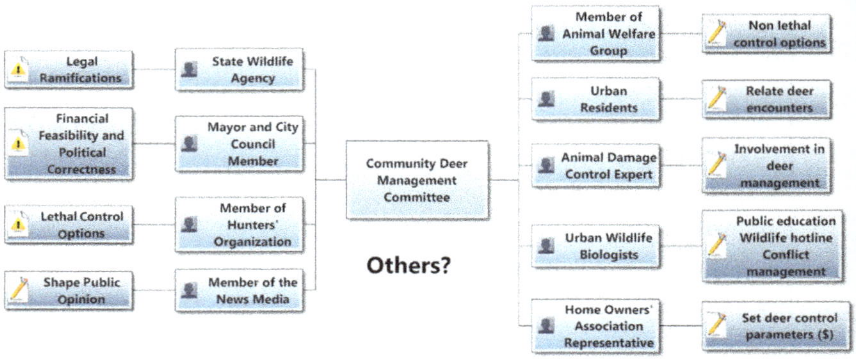

FIGURE 3.1 Illustration of some of the complexity of organizations and personalities involved in the formation of a community deer management committee (Adams 2016, Figure 14.5).

14. Urban wildlife management activities may not always be conducted by properly trained personnel, e.g., animal damage control, when compared to rural areas, e.g., private and state DNR wildlife biologists.

In summary, wildlife management is a challenging profession and process in urban habitats. The challenges have been discussed above in terms of comparisons of wildlife management in urban and rural habitats. Hopefully, this review enables readers to develop a deeper understanding of one of the many complexities of wildlife coexistence, i.e., deer management in their urban communities.

3.2 THE SCOPE OF THE PROBLEM

We were required to determine the magnitude of urban deer problems throughout the United States. This was one of the most challenging tasks in the production of this book. It became evident that urban deer management was not one of the top priorities of state DNRs or universities training future wildlife biologists (Adams 2003, study update: Sidebar 3.1). In fact, some states pay little attention to it! A survey conducted by the Quality of Deer Management Association (QDMA) asked state wildlife agencies if they had an urban deer management strategy in place and approximately how much time was allocated to urban deer issues. The evidence is undeniable; state agencies are not spending enough time on urban deer issues. In fact, according to the survey, only four states that responded to the survey (Maryland, Massachusetts, New York, and Illinois) spend 25% or more of their time on urban deer issues. Twenty of the 35 states (57%) that responded to the survey have an urban deer management plan in place (QDMA Staff 2016). However, urban wildlife represents one of the most pervasive, overarching, and universal wildlife management responsibilities in the entire United States (Adams 2016).

SIDEBAR 3.1

As illustrated in Tables 3.1 through 3.3, there are hundreds of urban areas (of those we could identify) that are faced with urban deer management problems. The scales of analyses included specific locations within cities, e.g., city parks, outdoor recreational areas, university campuses; cities within the state; state parks; suburbs and counties within metro areas; whole counties that were so heavily urbanized that every urban community within the county needed to be included, but were not identified by our sources of information, or the entire state had to be included. This made any analytical opportunity, e.g., city counts, impossible! Finally, for some states there was limited or no information on urban deer locations for any level of analysis.

The primary sources of information are listed as footnotes at the end of each table. Sources listed do not include the hundreds of websites required to gather additional information from newspaper articles, city council minutes on urban deer management, state DNR and city announcements of urban deer hunt locations, and state DNR and city urban deer management plans. Every state DNR, except for Hawaii (N = 49), was accessed online for the desired information. Some had *no* references to urban deer within their state. Others would reference urban deer in other states. It was a challenge to use the proper search terms or word combinations and to "force" specific state examinations. The reader may be interested in observing the results of an online search for "urban elk" as one illustration of assumed and anticipated results compared to actual results.

Only one professional journal article attempted to identify cities by state with urban white-tailed deer management issues (Conover 1995). It asked state DNR deer biologists to list urban locations of white-tailed deer populations within their states. Adams conducted an identical study in 2004 (unpublished) that also included mule deer, moose, and elk. Both studies revealed minimal recall by some respondents regarding locations of urban deer populations in their state. Tables 3.1 through 3.3 are the most comprehensive and contemporary analyses of urban locations of white-tailed, mule, and Key deer (n = 1,285), and moose (n = 140), and elk (n = 55) in the United States. We admit that even these results are "picking the low-hanging fruit" given the extensive distribution and large populations of deer in some states. As indicated earlier, a consistent scale of analysis, e.g., city, was impossible to achieve using the available information sources. Some respondents in both studies commented that every urban community in the state had deer populations. Others identified whole counties that were heavily urbanized with deer populations, e.g., Detroit Metro, including Macomb, Oakland, and Wayne Counties. In 2017, all three county governance units were planning a more aggressive urban deer bow hunt program, allowing ten females with one antlered deer after the female harvest. Chapter 2 explains the importance of concentrating on female rather than male harvests to expedite a successful urban deer management program. From the information given in these three tables, it is an intuitive assumption that the number of urban deer locations is proportional to the number of deer (Table 2.1) and their distribution in each state, and that appropriate urban deer management programs would be the result of these conditions.

A colleague inquired about the possibility of locating an urban community within a deer's national range that did not have an urban deer population. What conditions would preclude the presence of deer in an urban community? Where might these communities be located? This is a difficult question to answer given that white-tailed deer are found within the metro areas of New York, Minneapolis/St. Paul, Detroit, and others. An obvious answer would be any community where the resources for deer survival are not readily available, e.g., Zones T5 and T6 (see Figure 1.7 in a prior chapter). These zones typify loss of natural habitat due to the development of impervious surface covers and high density building complexes. This might include any inner city area that does not contain the required food, water, shelter, and human neighborhood support system for urban deer survival. It seems imperative to emphasize that deer and humans are highly adaptable even under such impoverished conditions.

3.3 POPULATION DYNAMICS

A finite world can support only a finite population.

Garrett Hardin

Population dynamics is the study of the factors that predict or explain a species' presence, abundance, and concentrations in a particular habitat. This section focuses on the population dynamics considerations of deer residing in urban communities. Population dynamics studies usually involve the following broad questions:

1. Can individual members of a population survive in the habitat being studied (e.g., an urban landscape)?
2. What adaptations do the species have that improve their chances of survival in this habitat?
3. What is the average density of the population for the species in this type of habitat?
4. What abiotic and/or biotic factors maintain and/or predict changes in the average population density over time?

The population dynamics of urban deer are influenced by human activity and development, including habitat fragmentation, the influence of suitable habitat corridors, supplemental feeding, animal damage control, and introduction of toxic materials into the urban environment.

3.3.1 How Populations Grow

The formula to calculate population growth is:

$$N_{t+1} = N_t + (B - D + I - E)$$

Where:

N = the population size at some point in time

t = time

t+1 = one time step into the future (e.g., next year)

B = number of births from *t* and *t+1*

D = number of deaths from *t* and *t+1*

I = number of immigrants from *t* and *t+1*

E = number of emigrants from *t* and *t+1*

The movement of animals in (*immigration*) and out (*emigration*) of the habitat is an important contributor to urban ecosystem population densities. Population size increases as a result of births and immigrants and decreases as a result of deaths and emigrants. The equation above is a mathematical way of saying that population change over one unit of time equals the population at the beginning unit of time plus the difference in births and deaths, plus the difference in immigrants and emigrants. Also, in the population ecology literature, the population growth rate (Figure 3.2) is referred to as the *intrinsic rate of increase* or *biotic potential.*

3.3.2 Population Growth Rate Patterns

The ways in which population densities change over time are illustrated in Figure 3.3. The J-curve demonstrates exponential population growth under optimal conditions. The S-curve shows a population at equilibrium but constrained by factors that limit additional growth. The J-curve, characteristic of some urban wildlife, represents a population explosion, e.g., moose population on Isle Royale in 2017. However, no population can grow exponentially for an indefinite period of time, for limiting factors place constraints on population growth. Members of a population will eventually produce more offspring than can be adequately supported by the urban ecosystem, also known as the *biological carrying capacity.*

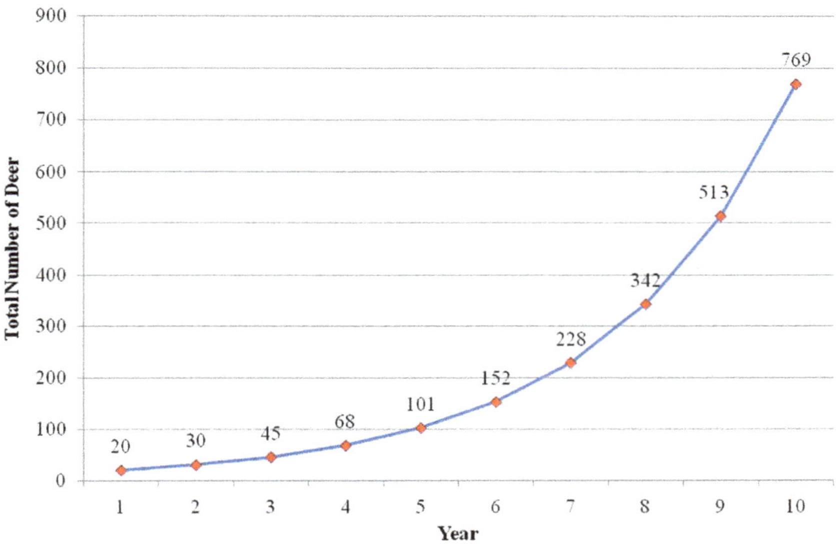

FIGURE 3.2 Example of potential deer population growth over a ten-year period. Here a deer population starting with 10 males and 10 females is projected for 10 years, assuming 50% fawn recruitment and 50% fawn mortality. A population beginning with a total of 20 individuals will grow to 769, degrading and exhausting habitat resources (courtesy Clark E. Adams).

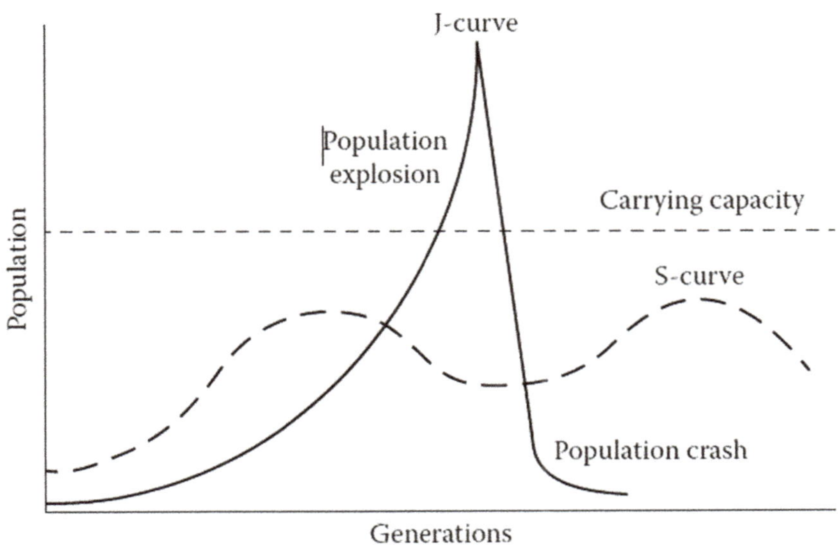

FIGURE 3.3 Population growth curves (courtesy Clark E. Adams and Cassandra LaFleur Villarreal).

For example, yearling white-tailed female deer generally produce a single fawn, but in areas with good forage, older does commonly produce twins or even triplets (DeNicola et al. 2000, see also Table 3.4). Each cervid species has unique reproduction characteristics as listed in Table 3.4. Although all the cervid species in Table 3.4 are polygamous, there are some variances among species. For example, moose bulls can mate up to three times per day ("Rutting Behavior of Moose"), while white-tailed deer bucks sometimes form a connection with a single doe and follow her for several days (Dewey 2003).

But as the deer population increases, plant resources become over utilized and the availability of good forage decreases to a level that will not meet the nutritional needs of all the herd members. This results in malnutrition or starvation, highly localized deer densities (indicated by high numbers of road kills), and the possibility of a massive population crash as illustrated in Figure 3.3. However, the urban ecosystem has buffers that limit the severity of a population crash. The human residents, by and large, find it difficult to tolerate the sight of starving deer and begin to provide additional food resources (e.g., corn), although the resultant larger congregations of animals facilitate disease transmission between individual animals, which may eventually bring the population back to carrying capacity levels.

Deer are the most recognizable and charismatic species of wildlife, but they are the cause of a growing urban wildlife management problem in many metropolitan areas throughout the United States. As urban sprawl increases, the natural habitat required by many wildlife species disappears, but deer are able to adapt to urban environments and related human activities. Deer populations grow rapidly in these areas due to (1) the lack of natural predators, (2) patchy habitats, (3) abundant food resources, and (4) increased offspring survival (Figure 3.4).

When humans and deer live close to one another, an increase in human/deer conflict occurs. Although deer are often viewed as an aesthetically pleasing addition to many homeowners in urban communities, they can cause several problems when they become overabundant (exceed the capacity of the native plant community) and unmanaged. For example, overabundant deer in urban communities can cause economic hardships by consuming landscape and garden plants, colliding with automobiles, which creates the need for repairs, and most importantly, bringing about human injuries and even fatalities, which results in added emotional costs.

A national survey, which included 204 respondents from state departments of natural resources, representing 40 states and the District of Columbia, was conducted in 2007 to assess the extent of human/deer conflicts throughout the United States. Eighty percent of the respondents reported that some of the communities in their state were experiencing

TABLE 3.4
Reproductive Characteristics of Cervid Species

	Moose (*Alces americanus*)	Elk (*Cervus canadensis*)	Mule Deer (*Odocoileus hemionus*)	White-Tailed Deer (*Odocoileus virginianus*)	Key Deer (*Odocoileus virginianus clavium*)
Breeding Season	September and October	Late September or early October	September–March	October–December	September–January (peaks in October)
Average Number of Offspring	1	1	2	2	1
Number of Offspring Range	1–2[a]	1–2[b]	1–4[b,c]	1–3[a]	1–2[b]
Gestation Period	8 months	8–9 months	6–7 months	6.5 months	6.5 months
Male Sexual Maturity Age	2 years	16 months	18 months	13–14 months	1.5 years
Female Sexual Maturity Age	2 years	16 months	18 months	10 months	7 months
Average Lifespan	8–12 years	20 years	7–10 years	2–3 years	6.5 years (female) 2.9 years (male)

(*Continued*)

TABLE 3.4 (CONTINUED)
Reproductive Characteristics of Cervid Species

Notes:

[a] Twins can be common, especially in more mature females.

[b] Twins are uncommon.

[c] Triplets and quadruplets are rare.

References

1. Dewey, Tanya, et al. "Alces Americanus (Moose)." Animal Diversity Web, animaldiversity.org/accounts/Alces_americanus/.
2. "Rutting Behavior of Moose (U.S. National Park Service)." National Parks Service, U.S. Department of the Interior, www.nps.gov/articles/aps-v5-i1-c7.htm.
3. Reed, Tom. "Hear the Bugle of Yellowstone Elk During Fall Rutting Season." My Yellowstone Park, September 12, 2018, www.yellowstonepark.com/things-to-do/hear-elk-bugling-rut-season.
4. Senseman, Rachel Lesley. "Cervus Elaphus (Elk)." Animal Diversity Web, animaldiversity.org/accounts/Cervus_elaphus/.
5. Ferguson, Kenneth I. Mule Deer (Odocoileus Hemionus). Mule Deer (Odocoileus Hemionus), NRCS, USDA, 2005. Fish and Wildlife Habitat Management Leaflet, Number 28
6. Odocoileus Hemionus, www.fs.fed.us/database/feis/animals/mammal/odhe/all.html#LifeSpanAndSurvival.
7. Dewey, Tanya. "Odocoileus Virginianus (White-Tailed Deer)." Animal Diversity Web, animaldiversity.org/accounts/Odocoileus_virginianus/.
8. Kime, Lynn. "White-Tailed Deer Production." Penn State Extension, August 3, 2019, extension.psu.edu/white-tailed-deer-production.
9. "Frequently Asked Questions about Key Deer – National Key Deer Refuge – U.S. Fish and Wildlife Service." U.S. Fish & Wildlife Service, www.fws.gov/Refuge/National_Key_Deer_Refuge/wildlife_and_habitat/KD_FAQs.Html#Q: How Long Do Key Deer Live?,
10. "Land Mammals – Key Deer." Florida Fish And Wildlife Conservation Commission, myfwc.com/wildlifehabitats/profiles/mammals/land/key-deer/.

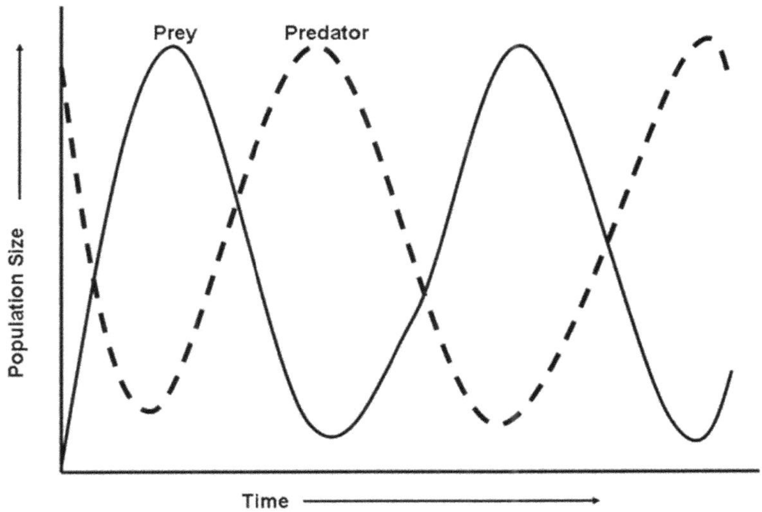

FIGURE 3.4 Predator and prey population relationship (courtesy Clark E. Adams).

human/white-tailed deer conflicts. Over the past ten years, most respondents (71%) reported an increase in the number of citizen and community complaints concerning human/white-tailed deer conflicts. However, less than half (42%) of the respondents could identify who was in charge of urban white-tailed deer management in their state. A respondent in Ohio suggested that any urban center in the state that had deer also had conflicts. As human populations continue to grow and communities expand, this sentiment will be reflected by residents of several communities within many states. The overarching conclusion from this study was that any urban community within the zoogeographic range of the deer in America will, in time, probably have problems with overabundant urban deer herds (Figures 3.5 through 3.8).

The dilemmas associated with high deer populations can become the burden of individual residents, homeowners' associations (HOA), and city governments. However, few are equipped with the knowledge to manage the problem. The goal of this publication is to educate readers on the biology and ecology of urban white-tailed deer and to provide the fundamental considerations required to develop an effective management plan.

3.4 HISTORY OF DEER ABUNDANCE

White-tailed deer have an important place in our nation's history. Venison and other products from deer provided sustenance, clothing, and even served as currency for early Americans. Today, most people fail to

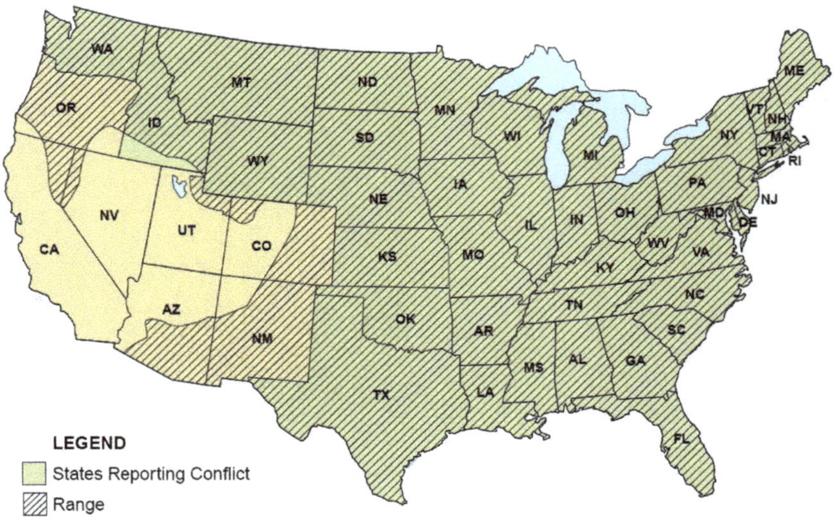

FIGURE 3.5 States reporting urban white-tailed deer conflicts with range overlay. Cathey, J. C., S. L. Locke, C. E. Adams, S. Ramirez, J. Alderson, and K. Schwausch. "Managing Overabundant White-tailed Deer." Texas A&M University, Texas AgriLife Extension Service, Texas AgriLife Research and Texas Parks and Wildlife Department, College Station and Austin, USA, 2010.

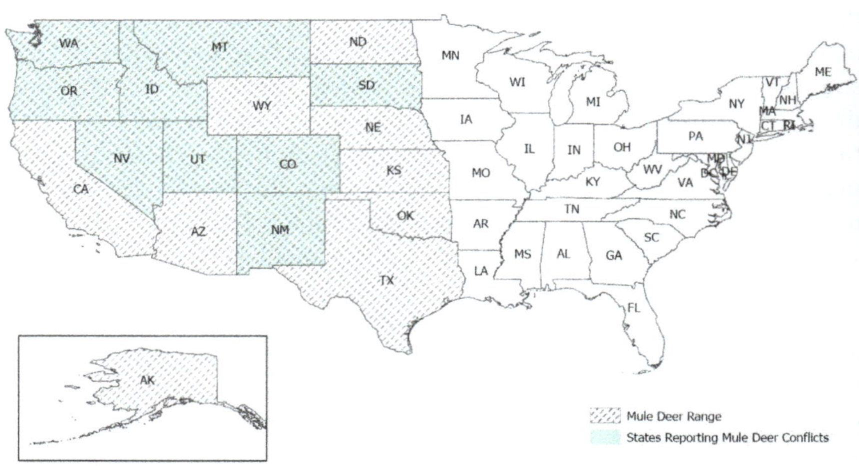

FIGURE 3.6 States reporting urban mule deer conflicts. Data for figure based on Adams, C.E. 2004. National assessment of urban deer (unpublished data).

recognize that white-tailed deer were once overhunted to the point where their very existence was threatened. Early on, white-tailed deer populations were not regulated by bag limits, and overhunting caused populations to decline. In the early 1900s, deer populations were at their lowest, which

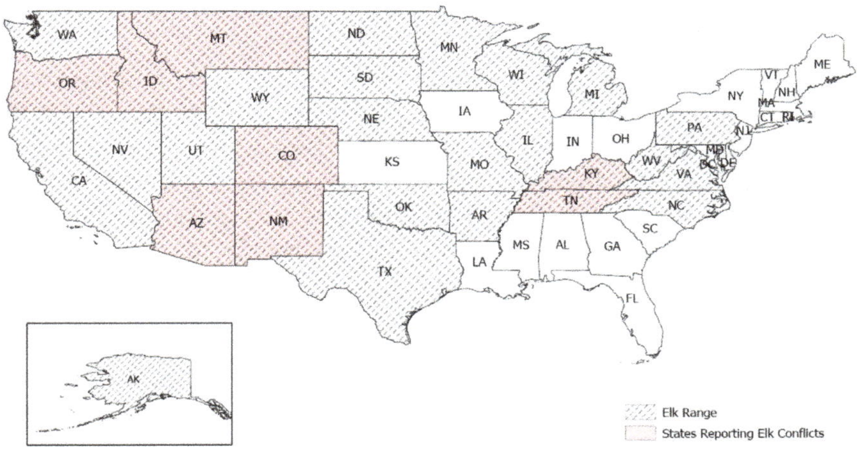

FIGURE 3.7 States reporting urban elk conflicts. Data for figure based on Adams, C.E. 2004. National assessment of urban deer (unpublished data).

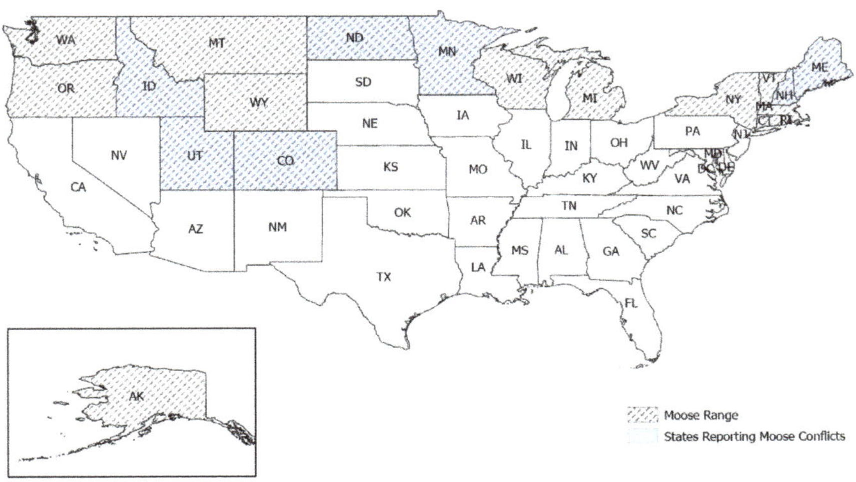

FIGURE 3.8 States reporting urban moose conflicts. Data for figure based on Adams, C.E. 2004. National assessment of urban deer (unpublished data).

coincided with early conservation work and the development of ideas that established wildlife management as a profession.

In Texas, deer populations have been influenced through an evolving process ranging from a hands-off approach to intensive habitat and population management. Today, management efforts are largely centered around producing quality deer found on farms and ranches in rural Texas. Managing deer at or just below the land's *biological carrying capacity* is a common goal. In rural areas, deer numbers can be managed by harvest to maintain a balance with available resources, but in urban locales, this task is more difficult.

In addition to deer, Texas also has an increasing urban human population that is predicted to swell from 23 million to 33 million by 2030. Consequently, as our population grows, the habitat for white-tailed deer is converted or lost completely through urbanization. Nevertheless, even dramatically altered landscapes like suburban yards are acceptable alternative habitats for white-tailed deer, and their populations often become quite large.

3.5 FACTORS LEADING TO DEER OVERABUNDANCE

White-tailed deer represent a conservation success story, in that their numbers rebounded from near extinction to the thriving populations we see today, particularly in urban neighborhoods. However, officers for homeowners' associations may view high deer numbers as a health risk based on diseases, deer-vehicle collisions, and other human/deer interactions.

Humans value white-tailed deer in dramatically different ways. On the one hand, there is the human value called the "Bambi syndrome" that is shaped by fanciful cartoons, where deer are cared for by Mother Nature without intervention by humans, especially hunters. Others view deer as ravenous eating machines spurred on by the need to reproduce and the need to be controlled. The fact remains that many urban communities are experiencing overabundant deer populations, urban sprawl, and limited natural resources. Consequently, some form of deer management becomes an absolute requirement, not an option.

3.5.1 Biological Carrying Capacity

Biological carrying capacity is a concept within wildlife science. It is defined as the maximum number of individuals that a given environment can support without detrimental effects. Biological carrying capacity for deer is a moving target in that it changes yearly and seasonally. For example, spring-time typically provides more vegetation than winter, thereby potentially supporting a larger number of animals. Biological carrying capacities for an area should be based on the time when resources are lowest (i.e., winter) versus when resources are at their peak. The concept of biological carrying capacity is widely used for deer populations in natural habitats; however, it should be noted that carrying capacity is more of a theoretical concept and less a measurable practicality. For deer populations in Texas, a general rule of thumb is one deer per 15–20 acres of useable habitat (Cathey et al. 2010).

A variation of biological carrying capacity is *cultural carrying capacity*, which is the maximum number of deer that society will accept within the area or the number of deer that can compatibly coexist with the local human population. The cultural carrying capacity can be exceeded without exceeding the biological carrying capacity for the area because humans have different tolerance levels of deer numbers. In most situations, however, both are exceeded and are correlated (Cathey et al. 2010).

3.6 PROBLEMS ASSOCIATED WITH DEER OVERABUNDANCE

3.6.1 Deer-Vehicle Collisions

Deer-vehicle collisions have steadily increased across North America over the last two decades. State Farm Insurance estimates that 2.4 million deer-vehicle collisions occurred between July 1, 2007 and June 30, 2009. This number is up 18.3% over five years earlier. The Insurance Institute for Highway Safety estimates that deer-vehicle collisions cause more than 150 fatalities per year, and the average property damage cost is $3,050. In Texas, deer-vehicle collisions have jumped 33% in less than a decade, and Texas drivers have a 1 in 416 likelihood of colliding with a deer over the next year (Cathey et al. 2010).

Deer-vehicle collisions are often prominent in and around urban areas with high densities of deer. Urban areas associated with rapid increases in human population growth and increasing road density are also susceptible to increasing deer-vehicle collisions. Although several techniques (reflectors, whistles, warning signs, and reduced speed limits) have been evaluated to reduce deer-vehicle collisions, none have been consistently effective (Cathey et al. 2010). It should also be noted that vehicle repair costs, on average, are higher when colliding with a larger cervid, like moose and elk. The risk of a deer-vehicle collision is also higher during periods and seasons of greater movement, like spring (Laliberte and St-Laurent 2020).

3.6.2 Disease Transmission

Transmission of disease between deer and humans is a major concern for many communities with overabundant white-tailed deer populations. White-tailed deer are a host for the black-legged tick (*Ixodes scapularis* and *Ixodes pacificus*). Black-legged ticks can transmit several tick-borne diseases. The most notable, Lyme disease, is primarily found in the northeastern and north-central United States in addition to some western states, including Texas (Figure 3.9). The life cycle of black-legged ticks is complex; however, tick larvae typically contract the disease when feeding on

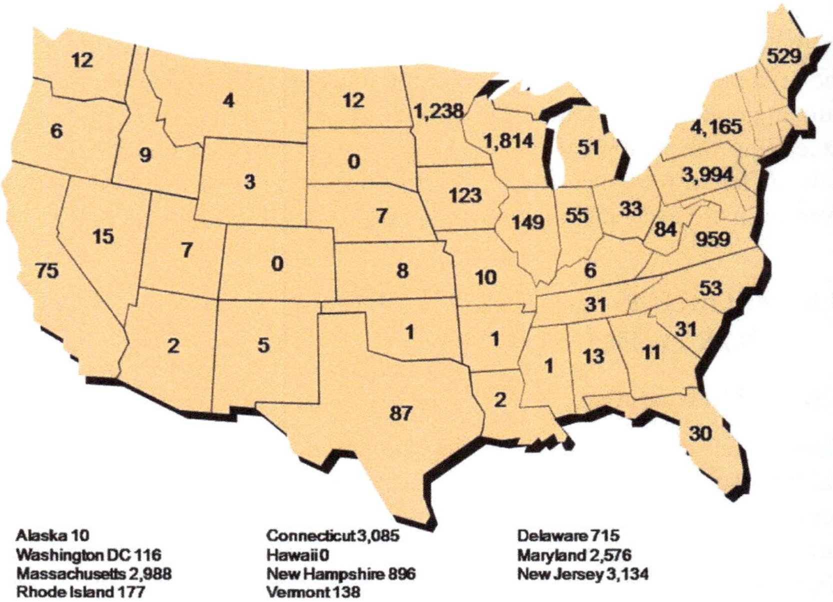

FIGURE 3.9 Number of reported cases of Lyme disease in the United States in 2007 (data from www.cdc.gov).

mice, birds, or other small animals. As adults, ticks can spread the disease to new hosts, which can include humans. Deer do not become infected with Lyme disease although deer are a major food source for adult ticks, and deer are important in transporting ticks and maintaining tick populations (Cathey et al. 2010).

According to the Center for Disease Control, approximately 21,000 cases of Lyme disease have occurred on average over the last ten years. Several studies have evaluated the impact of deer reduction on tick abundance. Despite conflicting results in various studies, a reduction in deer generally corresponded with a reduction of ticks. However, the level of deer reduction necessary to reduce the risk of Lyme disease to humans has not yet been established (Cathey et al. 2010).

3.7 ECOLOGICAL IMPACT OF URBAN DEER OVERABUNDANCE

Residents of communities with high deer densities recognize the adverse effect they have on gardens, flowerbeds, and ornamental landscaping. It has been estimated that deer cause approximately $250 million in landscape damages on an annual basis nationwide. Just as deer are capable of destroying ornamental landscapes, deer have a negative impact on native habitats where their numbers exceed the biological carrying capacity (Cathey et al. 2010).

By consuming leaves, stems, flowers, and fruits, deer directly affect the growth, reproduction, and survival of native plants. Many plants have evolved to cope with *herbivory* (consumption of a plant or its parts) through various means such as thorns and toxins. However, overabundant deer populations can consume large quantities of vegetation, creating a browse line or effectively consuming the vegetation within their feeding range (ground to 6 feet). A browse line or plants that show hedging are clear signals of high deer numbers. Deer will feed on preferred plants first, leaving plants with lesser wildlife value to dominate the area. Diversity of the overall plant community is reduced and regeneration of preferred plants is limited due to continued deer consumption (Cathey et al. 2010).

The impact of deer on the plant community has a cascading effect on other animal species and their own kind. Invertebrates, reptiles, amphibians, birds, and small and large mammals rely on diverse plant communities for food and cover. Recovery of habitat affected by deer is dependent upon several factors such as the intensity and severity of over-browsing, soil conditions, and climate. Some studies indicate that severe declines in deer density have little effect on plant diversity over the short-term and vegetation recovery may only occur through extended periods with low deer densities (Cathey et al. 2010).

3.8 LAWS AND REGULATIONS CONCERNING URBAN DEER

3.8.1 Who's in Charge of Managing Urban Deer?

Residents of communities with urban deer problems often ask, "Who is charged with managing urban white-tailed deer?" The difference between regulation and management becomes important in this discussion. Regulation is the process of developing laws that govern the well-being of selected animal species. Management is the manipulation of habitat and populations to establish a sustainable balance within the ecosystem. Responsibility for management is the same in rural and urban locations, which includes landowners, city leaders, and citizen organizations like homeowners' associations (Cathey et al. 2010).

3.8.2 Who's in Charge of Regulating Urban Deer?

In Texas, the Texas Parks and Wildlife Department (TPWD) is mandated by law to protect the state's natural resources. They do so by providing laws, regulations, permits, and educational materials to the public. The public can use these tools and recommendations in their decision-making process. It is the responsibility of communities to take action on management decisions and carry them out. This is the same strategy used on rural lands. The Texas Parks and Wildlife Department is not responsible for the hands-on management of white-tailed deer on private lands (Cathey et al. 2010).

3.8.3 Management Options in Texas

There are both nonlethal and lethal options for managing deer in an urban landscape (Table 3.5). Nonlethal methods do not reduce the population of deer. They modify the behavior of deer, restrict their movements and access to certain areas (e.g., roads, gardens), or encourage deer to leave the immediate area. Some of the nonlethal methods provide only a short-term fix (repellants fade quickly), while others can be a long-term solution (high fence can last many years) to human/deer conflicts. Regardless, nonlethal methods are best utilized in conjunction with a comprehensive deer management program (Cathey et al. 2010).

Fertility control is a popular choice among urban residents for managing overabundant urban deer herds. There is only one Food and Drug Administration (FDA) approved immunocontraceptive (GonaCon™) which is only available to federal and state wildlife agencies. It has not been approved for use in Texas. Fertility control does not reduce the number of deer; it only reduces their fertility, and it will not help in a situation where deer are already overpopulated. Fertility control is a viable option under certain conditions only, e.g., in a geographically closed population with a small number of deer. Another limitation of fertility control is the cost (ranging between $350 and $1,100 per treated female). In most cases, deer must be captured to administer the drug. Some fertility control agents are single dose drugs that last several years whereas others require annual booster shots to remain effective and thus require recapture of individuals annually. Time, labor, and equipment associated with deer capture are more costly than the drug (vaccine ranges between $20 and $100 per treated female). Studies indicate to maintain a population at a desired level then at least 50% of the females must be inoculated. Hands-on delivery of the drug is too cost prohibitive; consequently, passive delivery systems through supplemental feed must be developed

Lethal methods directly or indirectly reduce the deer population in one of three ways: (1) trapping and transporting deer to another location, (2) trapping and processing deer, or (3) harvest of deer through regulated hunting. Again, each method has advantages and disadvantages, and reduction methods often require an initial treatment phase (i.e., a certain proportion of deer are removed) followed by an annual maintenance phase (i.e., additional deer are removed). Maintenance phases may require a long-term effort to achieve and maintain lower population densities (Cathey et al. 2010).

TABLE 3.5
Comparison of Lethal and Nonlethal Techniques Used To Minimize Human/Deer Conflicts

	Method	Pros	Cons
Lethal	*Hunting*		
	Controlled hunting[a]: requirements and restrictions on who may hunt; hunting methods; hunting times and locations; and the sex, age and number of deer.	Only method with potential to generate revenue for landowners or communities. This method is inexpensive to communities and is considered an effective method. City ordinances may prohibit firearms thereby limiting hunting to archery only equipment.	Archery hunting method is less successful compared to firearms hunting. Lethal methods may not be approved of by many citizens, especially in dense communities. Individual hunting may cost between $85 and $300/deer.
	Traditional hunting[a]: licensed sportsmen and women using legal firearms or longbows.	Immediate removal and inexpensive.	Depending on state laws, citizens will need to be licensed. Although there have been no hunting accidents, safety is still a concern.
	Sharpshooting		
	Trained professionals used to harvest deer at short distances (<50 yards) [a].	Quick and discrete removal method.	Expensive, sharpshooter operations may cost up to $350/deer. Lethal methods may not be approved by many citizens.
	Trap and Euthanize		
	Deer are captured using box traps, clover traps, drop nets, or rocket nets. The deer can be euthanized by gunshot, penetrative captive bolt, or by pharmacological agent.	Preferred in areas where firearms are a major concern.	Not currently authorized by all state natural resource agencies. Cost of $175–300/deer.

(*Continued*)

TABLE 3.5 (CONTINUED)
Comparison of Lethal and Nonlethal Techniques Used To Minimize Human/Deer Conflicts

	Method	Pros	Cons
Nonlethal	*Habitat Alteration*		
	Unpalatable plants: *Flowers*: daffodils, chrysanthemums, oriental poppies, zinnias, and peonies. *Herbs*: garlic, chives, onion, lavender, sage, honeybush, and thyme. *Shrubs*: chokeberry, potentillas, barberry, bottle brush, rhododendron, wild Lilac, and honeysuckle. *Native plants*: black-eyed Susan, purple and prairie coneflower, eastern red columbine, butterfly weed, golden tickseed, yarrows, bergamot, and spiderwort.	To be effective, a majority of the vegetation must be removed. This method is more effective at low deer densities.	Plantings can become expensive, depending on the size of the area. Altering the habitat might impact native plants and other wildlife which many groups and regulatory agencies may oppose.
	Supplemental feeding	Can improve the condition of deer or take pressure off other food resources.	Expensive and deer become increasingly tame and more likely to tolerate human activity.
	Physical Barriers		
	High Fences (8 feet tall): Electric, woven wire, chain-link, stockade, PVC, rubber-coated chicken wire. Other: tree shelters	Effective method. Initial cost is dependent on materials used. Once installed, minimal maintenance is required. Over a 10-year period, a fence can save landowners from $10,000 to $25,000.	Some fences can be an aesthetic distraction. Fences work best for private property and do not solve a community-wide problem. In certain areas, electric fences are prohibited.

(*Continued*)

TABLE 3.5 (CONTINUED)
Comparison of Lethal and Nonlethal Techniques Used To Minimize Human/Deer Conflicts

Method	Pros	Cons
Deterrent Techniques[b]		
Visual: Lights, scarecrows, balloons, and shiny pans. Auditory: Noise makers. Olfactory: Commercial: Deer off, Deer Away, Tree Guard, Hinder, Detour, Hot Sauce, Ropel, and Orange TKO. Non-commercial: soap, human hair, and chemical mixtures. Dogs: used in a fenced area.	May be effective for short periods, but deer habituate quickly. However, the use of trained dogs is more effective than the other techniques.	Ineffective as deer become habituated to these techniques. As population densities increase, the effectiveness decreases. Many deterrents can be a nuisance to neighbors.
Predator Reintroduction		
Coyotes (*Canis latrans*), Bobcats (*Lynx rufus*), and Black bears (*Ursus americanus*).	Opportunity to return historical species. Minimal human involvement.	Predation is not sufficient to reduce high population densities. Not safe and is expensive.
Repellents		
Commercial: Deer off, Deer Away, Tree Guard, Hinder, Detour, Hot Sauce, Ropel, Orange TKO. Non-commercial: soap, human hair, and chemical mixtures.	Relatively effective at low deer densities.	Effectiveness decreases as deer densities increase. Can be costly and labor intensive. Chemical repellents may cause plant damage and leave noxious or offensive residues. Costs can range from $10 to $400 per acre.

(*Continued*)

TABLE 3.5 (CONTINUED)
Comparison of Lethal and Nonlethal Techniques Used To Minimize Human/Deer Conflicts

Method	Pros	Cons
Fertility Control		
SpayVac, GonaCon, Porcine zona pellucida (PZP): (1) contraceptive agents that prevent conception, and (2) abortion chemicals that terminate pregnancy.	Most common immunocontraception methods. Works best in a geographically closed population with a small number of deer. High success rates, 70%–90%.	Does not reduce the deer density; only reduces fertility. Costs per treated female vary from $350 to $1,100. Expensive treatments may require communities to raise money on their own.
Sterilization	Permanently prevents fawns from being born.	Deer are not removed from the population. Requires trained specialists and initial costs can be expensive, ranging from $800 to $1,000/doe.
Trap and …		
Relocate to the wild, deer farms, or adjacent properties.	Poses no risk to human safety and is a nonlethal solution to deer overabundance problems.	Ineffective and costly. Trapping operations can range from $150 to $500 per deer. Half of the deer captured and relocated die from capture-related stress. This can also contribute to deer/car collisions.
Passive		
Education, deer crossing signs, and supplemental feeding ban.	Inexpensive and increases the likelihood of viewing wildlife.	Communities must accept the consequences such as deer/car collisions and other damages.

(*Continued*)

TABLE 3.5 (CONTINUED)
Comparison of Lethal and Nonlethal Techniques Used To Minimize Human/Deer Conflicts

Notes:

[a] Bait stations can be used to improve shooter success.

[b] Motion activated sensors can be used in combination with other deterrent techniques such as sprinklers.

References

1. Bishop, P, et al. *A Citizen's Guide to the Management of White-Tailed Deer in Urban and Suburban New York*. New York State Department of Environmental Conservation, 2007, *A Citizen's Guide to the Management of White-Tailed Deer in Urban and Suburban New York*, hamilton-ny.gov/wp-content/uploads/2018/08/NY-DEC-Citizens-Guide-to-Deer-Mgt-07.pdf.
2. Cantu, Ruben, and Calvin Richardson. *Mule Deer Management in Texas*. Texas Parks and Wildlife, *Mule Deer Management in Texas*.
3. Cathey, J. C., S. L. Locke, C. E. Adams, S. Ramirez, J. Alderson, and K. Schwausch. Managing overabundant white-tailed deer. 2010. Texas AgriLife Extension Service Publication SP-38, Pp. 1–13.
4. *Chapter 3 – Moose*. Vermont Fish & Wildlife Department, pp. 40–52, *Chapter 3 – Moose*.
5. *City of Hendersonville, Tennessee Urban Deer Report*. Deer Advisor, 2014, *City of Hendersonville, Tennessee Urban Deer Report*.
6. Creacy, Greg. *Deer Management within Suburban Areas*. Texas Parks and Wildlife, 2006, *Deer Management within Suburban Areas*.
7. Dunn, John C. *City of Eau Claire Urban Deer Management Plan*. City of Eau Claire, 2015, *City of Eau Claire Urban Deer Management Plan*.
8. *Fertility Control and Mule Deer Population Management Fact Sheet #14*. Mule Deer Working Group – Sponsored by the Western Association of Fish & Wildlife Agencies, 2015, *Fertility Control and Mule Deer Population Management Fact Sheet #14*.
9. Kilpatrick, Howard J. and Andrew M LaBonte. *Managing Urban Deer in Connecticut – A Guide for Residents and Communities*. 2nd ed., Connecticut Department of Environmental Protection Bureau of Natural Resources-Wildlife Division, 2007, *Managing Urban Deer in Connecticut – A Guide for Residents and Communities*.
10. Rollins, Dale. *Predator Control as a Tool in Wildlife Management*. Texas A&M AgriLife Extension/Texas Sheep & Goat Predator Management Board, 2004, *Predator Control as a Tool in Wildlife Management*, agrilifecdn.tamu.edu/txwildlifeservices/files/2011/07/PredatorControlforWildlifeMgt.pdf.
11. Smith, Rosie, et al. *Hyperrabundant Moose Management Plan for North Mountain, Cape Brenton Highlands National Park*. Parks Canada – Unama'ki Institute of Natural Resources, 2015, *Hyperrabundant Moose Management Plan for North Mountain, Cape Brenton Highlands National Park*.
12. Stewart, C M. *Urban Deer Management: Characteristics, Options, and Studies*. Michigan Department of Natural Resources, 2016, *Urban Deer Management: Characteristics, Options, and Studies*.
13. *Urban Deer Technical Guide*. Indiana Division of Fish & Wildlife, *Urban Deer Technical Guide*.
14. *Utah Moose Statewide Management Plan*. Utah Division of Wildlife Resources- Department of Natural Resources, 2009, *Utah Moose Statewide Management Plan*.
15. Walter, W. D., Lavelle, M. J., Fischer, J. W., Johnson, T. L., Hygnstrom, S. E., & VerCauteren, K. C. Management of damage by elk (Cervus elaphus) in North America: A review. *Wildlife Research*, *37*(8), 2010, 630–646.
16. Walter, W. David; Lavelle, Michael J.; Fischer, Justin W.; Johnson, Therese L.; Hygnstrom, Scott E.; and Vercauteren, Kurt C., "Management of damage by elk (*Cervus elaphus*) in North America: a review," 2010, *USDA National Wildlife Research Center – Staff Publications*. 1346.
17. Warren, Robert. "2000 Annual Conference of the Society for Theriogenology." Society for Theriogenology, *Overview of Fertility Control in Urban Deer Management*, 2000, pp. 237–246.
18. Washington Department of Fish and Wildlife. Washington State Mule Deer Management Plan, Wildlife Program, Washington Department of Fish and Wildlife, Olympia, WA, USA, 2016, 144 p.

3.8.4 Laws Governing Trap, Transport, and Transplant (TTT) and/or Process (TTP)

The Texas Parks and Wildlife Department has established rules governing two management options known as Trap/Tranquilize, Transport, and Transplant (TTT) and Trap, Transport, and Process (TTP) (see Table 3.5). Specific language for each method can be found in the Texas Administrative Code, more specifically Title 31, Part 2, Chapter 65, Subchapter C, rules 65.101–115 (Cathey et al. 2010).

In order to use these urban deer control methods, the governing bodies (homeowners' associations, cities, and municipalities) must obtain a permit from the TPWD. If a landowner requests white-tailed deer for their property, the deer may be transported to their location, but only after meeting the TPWD guidelines. For the TTT method, before deer are trapped/tranquilized, the TPWD must inspect the release sites for suitable habitat; receive and approve a Wildlife Management Plan; and receive a Site Information Form for every release location. The state requires that 10% of deer be tested for CWD before they are relocated or processed. Processed deer meat is donated to charitable organizations like Hunters for the Hungry (Cathey et al. 2010).

3.8.5 Hunting

Hunting is an effective method for controlling overabundant urban deer. Hunting can be conducted by licensed hunters or professional sharpshooters. It is critical that deer harvest, through hunting in urban communities, is conducted to maximize human safety, humaneness, discretion, and efficiency. City ordinances may prohibit the use of firearms, thereby limiting hunting to archery only equipment (Cathey et al. 2010).

In communities where hunting is feasible, hunters must meet proficiency standards established by the community and state game agency in order to participate. Hunters must be evaluated to determine that they are responsible and ethical. Common areas within the community are located to maximize human safety and efficient deer harvest. Deer are often baited into the area for a period of time, and then hunters are placed at those locations. Meat from carcasses must be used by the hunter or donated to a designated charitable organization. Both hunters and non-hunting residents must recognize that this is a management hunt designed to reduce the number of deer, most likely by removing females. This is not a trophy buck hunt (Cathey et al. 2010).

Sharpshooting uses trained professionals (local policemen and women) to harvest deer. Sharpshooters use baited sites to attract deer and take relatively short shots (<50 yards). Shooting is often done from an elevated position so that the shot's trajectory is toward the ground. Head and neck shots ensure a quick death. Sharpshooters typically use sound suppression devices on their firearms to limit noise and disturbance. Sharpshooters often work at night and are capable of removing many deer as compared to daytime hunting by residents who are limited by the number of license tags (Cathey et al. 2010).

3.9 THE TALE OF TWO CITIES

3.9.1 Background

Two Texas communities were chosen to determine what residents knew about the ecology and management of urban white-tailed deer. Lakeway, Texas, is located northwest of Austin, and Hollywood Park, Texas, is located in the north-central area of San Antonio. Based on the U. S. Census Bureau's 2015 estimates, San Antonio is the 2nd largest city and Austin is the 4th largest city in Texas. In addition, San Antonio was the fastest growing city in the United States in 2016 and 2017. Both communities have similar geographic, ecological, and socioeconomic characteristics, which include:

1. High density of urban white-tailed deer populations.
2. Relatively abundant green space and lot sizes.
3. Potential for white-tailed deer population growth.

Since the 1990s, urban deer herds have negatively impacted Lakeway and Hollywood Park through increased deer-vehicle collisions, defined browse lines throughout the community, and human-deer encounters. Each community has a Deer Control Committee that decides, in cooperation with the TPWD, strategies for managing the urban deer herds. The success of the urban deer management plan in Lakeway and Hollywood Park is determined by changes in the number of deer-vehicle collisions, human-deer encounters, deer found injured or dead, and fence-related accidents. Strong differences of opinion between stakeholder groups in Hollywood Park have led to a more contentious urban deer management environment than is the case in Lakeway (Cathey et al. 2010).

Lakeway and Hollywood Park have implemented a city ordinance which prohibits the feeding of deer, restricts public access to deer controlled areas, and assigns penalties for damage or destruction of deer control equipment. Violators are charged with a Class C misdemeanor punishable by a fine of $1 to $500 if enforced by the local police department. These laws have created conflicts between residents and deer managers (Cathey et al. 2010).

3.9.2 Need for Educational Program Development

A series of 16 questions was used to determine what Hollywood Park and Lakeway residents knew about urban deer ecology and management (Table 3.6). Scores ranged from 5 to a perfect score of 16. The average score for residents of both communities was ten correct answers. The majority were unaware of how urban deer feed, breed, and die, and the consequences and effectiveness of two management options. A 16-point examination does not provide all of the information required for developing comprehensive educational programs on urban deer ecology and management, but it did indicate a need to develop educational programs on:

TABLE 3.6
Survey Response by 256 Residents of Hollywood Park and Lakeway, TX, Concerning Urban Deer Ecology and Management (2008)

Statements	Answer	(%) Correct
1. The size of urban habitats prevents deer herds from growing too large.	False	60
2. Urban deer can carry diseases that affect humans.	True	59
3. Urban deer can destroy habitats used by other animals.	True	52
4. In order to survive, urban deer rely heavily on supplemental food sources.	True	36
5. Urban deer can begin to produce fawns when they are six months old.	True	17
6. Twins are a common result from urban deer reproduction.	True	52
7. More deer per acre means the deer will be physically larger in size.	False	86
8. The most common cause of death in urban deer herds is predators.	False	75
9. The least common cause of death in urban deer herds is disease and starvation.	False	49
10. Deer are an endangered species in Texas.	False	91
11. Deer live in urban areas because they have adapted to living near people.	True	55
12. People can help urban deer the most by letting nature take its course.	False	41
13. Deer live in urban areas because human development has pushed them out of their natural habitat.	True	81
14. People have done more harm than good for urban deer.	True	60
15. People can help urban deer the most by managing them.	True	69
16. Fertility control techniques for managing urban deer are cost-effective and easy to implement.	False	10

(1) Deer population control methods.
(2) Basic deer biology and ecology.
(3) Management alternatives.
(4) Long-term strategies to solve human-deer conflicts.

Today, Lakeway deals with less controversy from its residents regarding urban deer management practices than in the 1990s when deer management first began. However, Hollywood Park continues to struggle with protesting residents who are opposed to deer management. The current deer management programs in Lakeway and Hollywood Park seem to be effectively managing the deer population. Lakeway is currently using the TTP method and Hollywood Park is practicing the TTT method to manage urban deer in their communities. In the past, Lakeway and Hollywood Park have removed in excess of 250 deer a year in order to sustain a healthy deer population and ecosystem within their cities (Cathey et al. 2010).

Often urban residents do not recognize that if deer management were to stop, populations would once again exceed the biological carrying capacity. Residents would then experience more economic loss and increased deer-human encounters. It is important for managers to understand that when dealing with the public, a variety of issues will arise from different stakeholder groups. It is good to address issues before management programs are implemented in order to reduce the potential conflict between managers and residents (Cathey et al. 2010).

4 Urban Deer Management Planning
Process, Progress, and Problems

Urban deer management can be defined as the manipulation of deer populations and habitat to achieve a goal.

paraphrased from Sargent and Carter, 1999

Urban deer management is not just about deer. There are three inseparable components that include deer, humans (cultural, social structures, and institutions), and habitat (Krausman et al. 2013). This may be called a holistic approach to human manipulation of a particular wildlife species that takes into consideration all relevant components involved in the process. The authors were challenged to find published urban deer management research that utilized this comprehensive approach. More often than not, single issues (e.g., deer, human dimensions, deer habitats) were the focus. For example, these were a couple of the top results when searching for "urban deer management" journal articles:

- Messmer, Terry A., et al. "Stakeholder Acceptance of Urban Deer Management Techniques." *Wildlife Society Bulletin* (1973–2006), vol. 25, no. 2, 1997, pp. 360–366.
- Nielsen, C., Anderson, R., and Grund, M. Landscape Influences on Deer-Vehicle Accident Areas in an Urban Environment. *The Journal of Wildlife Management*, vol. 67, no.1, 2003, pp. 46–51.
- Kilpatrick, Howard J., et al. "Assessing Strategies to Improve Bowhunting as an Urban Deer Management Tool." *Wildlife Society Bulletin*, vol. 32, no. 4, 2004, pp. 1177–1184.

4.1 STEPS FOR MANAGING URBAN DEER POPULATIONS

There are many communities throughout the United States that are in one of three stages in the urban deer management process. These three stages can be classified as:

1. Emerging: Deer are present, and there are beginning signs of browsing on landscape plants and garden vegetables. The deer population is increasing each year, but incidences of human/deer conflicts are still low. This is the ideal time to initiate a proactive urban deer management program before the next stage becomes evident.
2. Crisis: The problem of a growing urban deer herd has been ignored for so long that the community is literally overrun with deer, e.g., one deer per acre. There are noticeable browse lines in the trees, hedging of shrubs, and loss of understory plants. Incidences of human/deer conflicts are high with exceptional loss of landscape plants and garden vegetables and increased deer/car accidents. The deer herd has a noticeable number of stressed deer from disease or starvation. Development of an urban deer management plan is no longer an option, but an absolute necessity.
3. Managed: These are communities that have initiated an urban deer management program by completing the list of steps discussed below. In fact, urban deer management is an ongoing process where a pre-determined number of deer are removed from the community on an annual basis.

There are several sequential steps that need to be taken in order to develop an effective urban deer management program.

Step 1: Determine the Scope of the Problem

This is a relatively easy process, relying on the residents' powers of observation and the responses to an online citizens' survey (Appendix). Community residents are the best informants when it comes to providing information on the number of deer and their impacts on the community. The citizen survey also provides an opportunity to determine community residents' opinions on deer numbers, desire to implement management options, feasible and acceptable management options, and who is in charge. The citizen survey can also determine the level of knowledge that residents have about urban deer ecology and management options which provides guidance in the development of relevant educational materials. It is strongly suggested that communities seek out survey research professionals to design, administer, analyze, and report the results of citizen surveys. Previous uses of citizens' surveys in urban communities with overabundant urban deer herds have revealed one overriding and common trend in the results. *Urban deer management is more a function of people management rather than deer management.* Once Step 1 has been completed, the community can move to the second step if justified from the results of the citizens' surveys.

Step 2: Form a Deer Management Action Committee

Some community entity, usually the city council, mayor's office, or homeowners' association, has to put out an announcement that there will be a formation of the Deer Management Action Committee or a Deer Advisory Committee, invite residents to become members, and attend an organizational meeting. Their first order of business is to review the results of the citizens' survey of the community, invite the professional input of TPWD urban wildlife biologists, and organize a town hall meeting.

Step 3: Have a Town Hall Meeting

The first consideration in Step 3 is to identify the relevant stakeholders that might participate. Some examples are community residents, city and state government, homeowners' associations, law enforcement, animal welfare groups, media, hunter organizations, and others that might be identified from the results of the citizen's survey conducted in Step 1. The second consideration is, who will call the meeting to order? Some previous examples have been homeowners' associations and city councils. It should never be the regulatory agency, e.g., TPWD, that chairs the meeting! TPWD has its own unique role in urban deer management as discussed above. It is during the town hall meeting that the results of the citizen's survey will be presented. Additional information may be provided by TPWD urban wildlife biologists. Expect the first meeting to be a venting and even a contentious session, given the attendance of residents who are at totally opposite ends of the deer management spectrum. The whole purpose of the town hall meeting(s) is to provide as much relevant information as possible so that the citizens can participate in the development of an urban deer management plan that is acceptable and, more importantly, effective.

Step 4: Develop an Urban Deer Management Plan

Prior to developing an urban deer management plan, it is instructive to understand those factors that contribute to deer abundance in North America. Deer are superbly adapted to exploit the resources in urban areas and can easily develop exponential (e.g., 2, 4, 8, 16, 32, 64, etc.) growth patterns because of the following conditions:

1. Low abundance of natural predators. Large predators are the first species eliminated during urban sprawl.
2. Healthy breeding habitat conditions that include patchy and defined edge habitats.
3. Tolerance of urban disturbances including human presence and their activities.

4. Longer life spans in the city when compared to the country.
5. Lack of hunting in urban areas and regulatory protections at the state level.
6. High production and survival rates of offspring.
7. Abundant alternative food resources in the form of ornamental shrubs, garden plants, succulent grasses, small plants, and supplemental feed.

In general, management plans should be focused on reduction and control of deer populations and to reduce related damage. Different communities will need different plans that best fit existing conditions. There are four urban deer management approaches:

Avoiding the problem: This is a proactive management strategy that involves urban development that integrates rather than excludes nature. It involves community designs that preserve the natural habitat, are sustainable in the use of natural resources, and reconnect human society with the natural world they live in. Smart Growth, Design with Nature, and Conservation Design are three examples of existing urban development alternatives that avoid the problem of deer (wildlife) overabundance in urban communities.

Getting at the root cause: Getting at the root causes of deer overabundance in urban areas requires an analysis of factors (see the above list of seven factors) that promote deer presence in a "typical" urban community. The most common management action required to get at the root cause is "to clean up the neighborhood." For example, remove all supplemental feeding programs and impose heavy fines on those residents who refuse to stop feeding the deer, and use landscape plants unpalatable to deer. The only predatory pressure that can be applied would include trapping and hunting as discussed earlier.

Attack the symptoms: The most prevalent management strategy to reduce or control white-tailed deer overabundance in urban communities does little more than attack the symptoms, i.e., a reactive management strategy. The cliché "winning the battle, but losing the war" is used appropriately in this case. Management strategies that attack the symptoms include those that (1) clean up the mess – feces and automobile wrecks, (2) cull the herd of some members, (3) move the problem elsewhere – translocation, (5) high or electric fences, (6) behavioral modification – aversion techniques, (7) introduce pseudo predators, e.g., trapping and hunting, and (8) fertility control or contraceptives. These strategies are usually applied to get short-term results in the most problematic areas.

Do nothing: There are groups of stakeholders that advocate "live and let live (or die)" or "let nature take its course" philosophies of management regardless of population size. Advocates of this urban deer management option will need to prepare themselves and other community residents with an increased loss of vegetation, deer/vehicle accidents, deer starvation (emaciated animals are a pitiful sight), and a loss of species diversity in the community. A do-nothing approach to urban white-tailed deer management will probably bring little satisfaction to anyone as a credible approach to the problem.

Step 5: Implement the Plan and Monitor Results

Most communities have adopted urban deer management plans that are a combination of getting at the root cause and attacking the symptoms. First, a credible method of estimating herd size needs to be applied. The TPWD urban and deer biologists may be called in to perform this function. Otherwise, the job can be given to a private contractor. Next, a professional trapper, with experience and TPWD regulatory clearances, is hired to remove the excess number of deer determined from the credible estimate. In most cases, the recommended number of deer per acre of an urban community is 1 deer per 15 acres. The effectiveness of the management plan may be monitored in terms of reduced plant damage, deer-vehicle collisions, loss of deer from disease and starvation (the deer herd will appear healthier), and human/deer conflicts. However, expect annual additions to the deer herd from prolific reproduction and possible immigrations. This also needs to be monitored closely. The bottom line for communities that have not adopted the avoiding-the-problem approach to urban deer management is to accept the fact that their plan will be an ongoing obligation as long as there are deer in the neighborhood. The case study that follows provides evidence of this management obligation.

One hundred and three urban deer management plans from 44 states were reviewed. Organized special hunts were one of the most common (76%) means of urban deer control. Sharp-shooters were incorporated into 72% of management plants, followed by the use of high fences and landscape alteration in 68% and 64% of the plans, respectively. Repellents were used in 60% of the management plans and only 28% of plans contained a public education component. Trap/Tranquilize, Transport, and Transplant (TTT) and Trap, Transport, and Process (TTP) methods of urban deer population control were mentioned in 12% and 32% of the plans, respectively. Only 2% of the management plans used fertility as an urban deer control measure.

The array of options to estimate urban deer numbers included spotlight counts (54%), check stations (62%), and aerial surveys (52%). Some of the management plans also used thermal imaging (14%), camera monitors (8%), harvest surveys (8%), and pellet counts (6%).

The utilization of urban deer management plans falls under the responsibility of several parties. The state parks and wildlife or department of natural resources assume responsibility for 77% of the urban deer management plans. Other management plans are the responsibility of the city (35%), the communities (27%), and the county (13%). Property owners (2%) are also looked upon to be responsible for their own deer management.

Urban deer management plans can be utilized on many levels. Although many states are currently experiencing urban deer problems, only 37% of the management plans are implemented on the state level. Forty-eight percent of all the management plans reviewed were created and managed on a county, city, or community level. Ten percent of the plans reviewed did not specify if it was a state management plan or an individual county, city, or community plan.

Urban deer management costs money! Most (43%) of the urban deer management plans reviewed were state funded. Cities, counties, and communities were also responsible for funding their urban deer management plans. Cities and counties were the primary funding agents for 16% of their urban deer management plans while communities were responsible for funding 15% of the management plans. Of the plans reviewed, 34% gave no information on how their program is funded.

4.2 SUBURBAN WHITE-TAILED DEER MANAGEMENT

In the last 25 years, there has been a growing trend toward community-based management. Communities across the nation have been faced with resolving their own wildlife-related problems due to regulatory agencies focusing their efforts on other issues due to funding and agency priorities. As mentioned earlier, regulatory agencies should only provide subject guidance. A concept known as "co-management" consists of a partnership between stakeholders and community members who share responsibility and negotiate on behalf of the community (Curtis et al. 2000).

One method of co-management is a Citizen Task Force (CTF). A Citizen Task Force is a volunteer group that represents a broad range of stakeholders. The stakeholders can include a broad range of interested groups such as farmers, hunters/sportsmen, foresters and gardeners, conservationists, local officials, etc. By including a diversity of stakeholders, the community will have several voices and local knowledge will be represented. The roles of a CTF are to make favorable decisions for the community, raise awareness of the issue(s), serve as community advocates, facilitate communication between different agencies and stakeholder groups, set deer population goals and objectives, and finally, provide management recommendations (Curtis et al. 2000). In addition, CTFs can also create and distribute citizen/resident surveys and conduct deer counts. For example, the New York State Department of Environmental Conservation has utilized CTFs since the 1990s. Biologists provide the CTF optimal deer population ranges, and the task force members determine what the appropriate deer population range should be for their community (Frye 2006, "Public Input on Deer Population Size," Conover 2012). In many large communities or communities that have several interested members, elections or appointments may be necessary. Although there are several benefits of co-management, the community may not be entirely accepting of their recommendations, such as for bait and shoot.

The need for public involvement when developing community deer management programs is essential for a successful management plan. Integrating proven techniques into a long-term strategy will be more successful than seeking simple, quick fixes. Although nonlethal methods may reduce problems at specific sites, lethal population reduction programs usually are required to resolve community-wide conflicts. Because the reproductive output of deer that live in urban environments is commonly high, a decision to postpone active management often will lead to greater difficulty when efforts are implemented in the future (DeNicola 2000).

4.3 DRIVING FACTORS

Before communities even begin to consider developing a deer management plan, there are driving factors that lead to this phase. Many communities are faced with maintaining and reducing the deer population over an extended period of time, and the population can easily double without the right control method (1,000 deer over 10 years without control would become 20,000 deer).

Although hunting is the most practical and cost-effective means to control free-ranging deer populations in most settings, many residents are not convinced that it is a safe option for their community. Alternatives to hunting typically are limited in applicability, prohibitively expensive, logistically impractical, or technically infeasible (*An Evaluation of Deer Management Options* 2009). However, city ordinances and state laws must be taken into account when considering management options. Aside from the extreme cost, many states have not approved contraception methods, and these methods only reduce fertility, not the number of deer in the community. Another unpractical option, Trap/Tranquilize, Transport, and Transplant (TTT), requires regulatory permits, and the release sites must be approved by the regulatory agency.

The process that community residents need to go through to initiate and implement a deer management plan can take several years. For example, a community in Maryland, Lake Linganore, began exploring deer population reduction methods in September 2015; however, the bow hunts were proposed for January 2017, which was 16 months later. The deer management plans must be drafted and must go through a public comment period where residents can express their concerns, and finally, the plan must be approved and must move into the implementation phase. However, opposed residents will even file a petition if they disagree with the selected management option. For example, residents of an HOA community in Georgia, Whitewater Country Club, filed a petition online titled "Stop the White Water Creek HOA Board from 'eradicating' the deer" (Figure 4.1). Available technologies and procedures to "manage" urban deer populations are well established, but implementation in urban communities is the greatest deterrent to success. Each community will face its own set of issues, oppositions, and implementation times.

Driving factors can be categorized into the following four categories: ecological, cultural, economic, and political (Table 4.1). Ecological

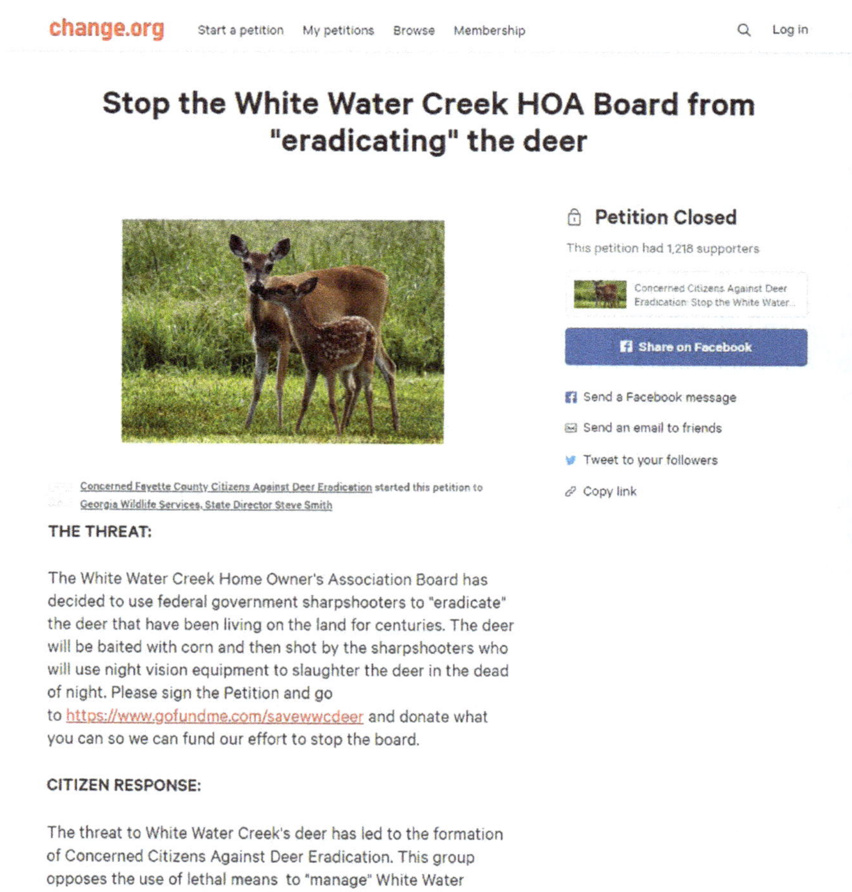

FIGURE 4.1 Residents of an HOA community in Georgia, Whitewater Country Club, filed a petition online titled "Stop the White Water Creek HOA Board from 'eradicating' the deer."

TABLE 4.1
Driving Factors That Ultimately Lead To an Urban Deer Management Plan

Ecological Issues	Cultural Issues	Economic Issues	Political Issues
1. Increased population densities: deer longevity and fertility rates within a confined space	1. Variations in public opinion on management options	1. Losses of landscapes and gardens	1. Legal considerations in urban deer management
2. Habitat loss due to overgrazing	2. The number of deer	2. Deer/car collisions including the death of both deer and humans	2. Negotiating compromise among heterogeneous stakeholder groups
3. The health of the deer population	3. Disease transmissions	3. Habitat alterations to exclude deer	
4. Lack of natural predator/prey relationship	4. Citizens Task Forces to deal with deer management		

driving factors include the biological carrying capacity, geographic distribution, habitat conditions – particularly ecological impacts on their habitat, an abundance of predators, and overall deer health. Cultural driving factors, on the other hand, are based on the human perspective, public opinions, and citizen comfort levels. Examples of cultural driving factors include cultural carrying capacity, public opinions on management options, diseases that could affect humans – Lyme disease – and creating and managing volunteer groups such as Citizen Task Forces and Deer Advisory Committees. Economic driving factors are those that are associated with any costs to homeowners, cities, and municipalities, etc. These costs can be from landscape damage, deer/car collisions, insurance rates, habitat alterations, deterrents, and even costs associated with implementing lethal and nonlethal methods. For example, if a community decides to move forward with a fertility control technique, who will be responsible for the cost? Finally, political driving factors are legal considerations and negotiations among stakeholder groups. Legal considerations can include state requirements and processes for permits, approved techniques (i.e., fertility control), HOA rules, creating volunteer groups – who gets to participate? Keep in mind, more often than not, that there will be a combination of these four categories (Figures 4.2 through 4.4, Tables 4.2 and 4.3).

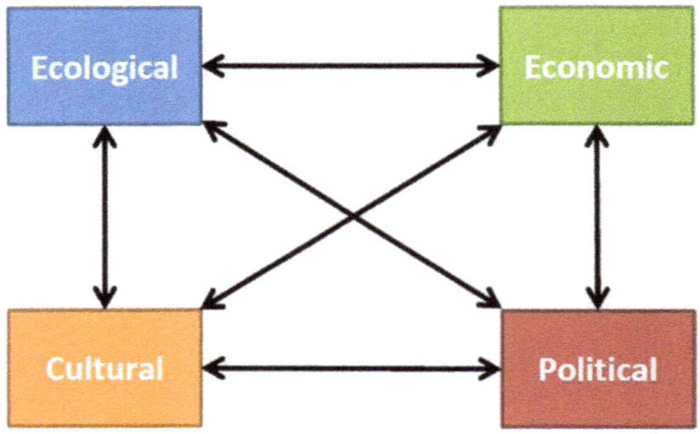

FIGURE 4.2 Driving factors are all connected.

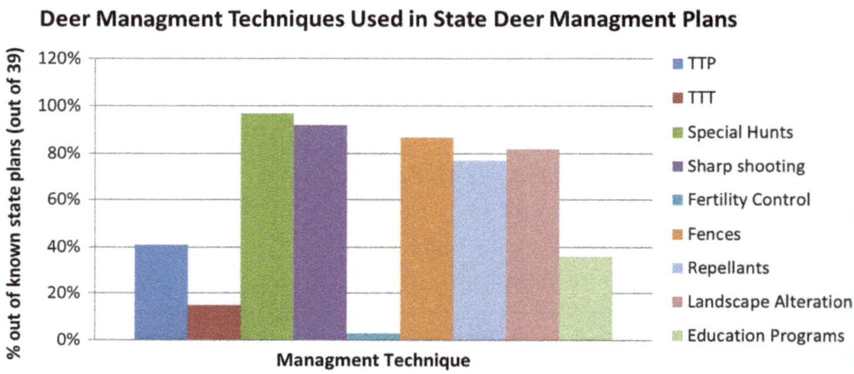

FIGURE 4.3 Management techniques percentages mentioned in state management plans.

4.4 WHO ARE YOU GOING TO CALL?

As a study for this book, we researched companies and agencies that could be contacted to help solve an urban community's overabundant deer population. This was a difficult task since there are not any published sources of information on the topic. In addition, the standard wildlife operator does not attend to urban deer concerns. The target "nuisance" species usually consists of insects and rodents.

We researched hundreds of companies by using key words such as "urban deer control," "deer population management," and even "urban deer removal." As an experiment, we also attempted to search companies by certain states such as Texas, Michigan, and Georgia. The search results were astounding. A majority of companies focus on methods that only attack the symptoms of the issue, particularly deterrents and fences. The

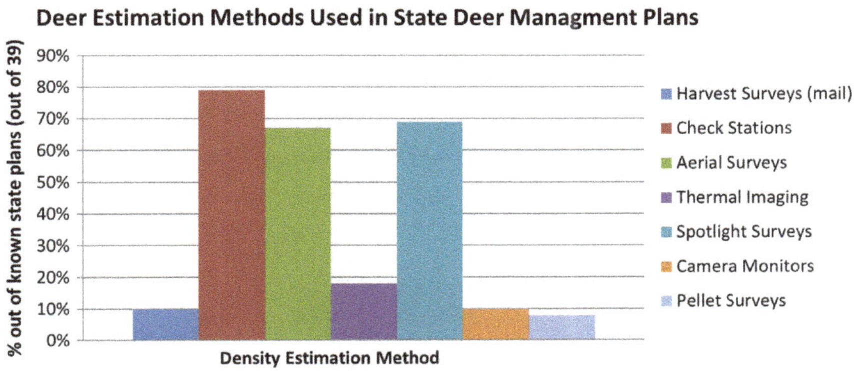

FIGURE 4.4 Census techniques percentages mentioned in state management plans.

most probable reason for companies offering only these approaches is the steps and costs that are involved with other approaches like TTT, TTP, and even fertility control.

During the search, we also discovered that some smaller cities, like Pierre, South Dakota, list the local police department or animal control department as responsible for any wildlife problems. (*Deer Management – Citizens Task Force* (Pierre, SD)). Many cities and private companies offer deer and wildlife removal services if a deer was hit by a car or died of natural causes. In Oregon, the management of deer populations is the exclusive purview of the Oregon Department of Fish and Wildlife. The Washington Department of Fish and Wildlife listed 556 wildlife control companies, but none specifically listed deer as a controlled species, although deer may have been implied under the broad category of "all authorized species" ("Hiring a Wildlife Control Operator"). South Carolina DNR listed 1,138 wildlife control companies, but none listed deer as a controlled species ("Wildlife Control Operators List"). People who specialize in handling wildlife damage problems can have different titles.

- NWCO (pronounced "newco"): Nuisance Wildlife Control Operator
- WCO: Wildlife Control Operator

Table 4.4 lists over 30 companies we found that mention *deer* as a target species. However, only half of the listed companies actually "manage" deer. Maybe the demand for companies to "manage" deer is not there. There might be several others, but they are not advertised on their website. Some companies list the contact information of certified wildlife biologists

TABLE 4.2
Deer Management Techniques Practiced by States

Management Technique	States That Practice Management Technique
TTP	Colorado, Idaho, Maryland, Minnesota, Mississippi, Montana, Missouri, North Dakota, New Hampshire, New Jersey, New Mexico, New York, Ohio, Texas, Virginia, Wisconsin
TTT	Idaho, New Jersey, New Mexico, Texas, Virginia, Wisconsin
Special Hunts	Alabama, Arizona, Arkansas, Colorado, Connecticut, Florida, Georgia, Iowa, Idaho, Illinois, Indiana, Kansas, Kentucky, Maryland, Maine, Michigan, Minnesota, Mississippi, Montana, Missouri, North Carolina, North Dakota, Nebraska, New Hampshire, New Jersey, New Mexico, New York, Ohio, Oregon, Pennsylvania, Rhode Island, South Dakota, Tennessee, Texas, Virginia, Vermont, Washington, Wisconsin
Sharpshooting	Alabama, Arizona, Arkansas, Colorado, Connecticut, Florida, Georgia, Iowa, Idaho, Illinois, Indiana, Kansas, Kentucky, Maryland, Maine, Michigan, Minnesota, Mississippi, Montana, Missouri, North Carolina, North Dakota, Nebraska, New Hampshire, New Jersey, New Mexico, New York, Ohio, Oregon, Pennsylvania, South Dakota, Texas, Virginia, Washington, Wisconsin, Wyoming
Fertility Control	New York, Pennsylvania
Fences	Alabama, Arkansas, Colorado, Connecticut, Florida, Georgia, Iowa, Idaho, Illinois, Indiana, Kansas, Kentucky, Maryland, Michigan, Minnesota, Mississippi, Montana, Missouri, North Carolina, North Dakota, Nebraska, New Jersey, New Mexico, New York, Ohio, Oregon, Pennsylvania, Rhode Island, South Dakota, Texas, Virginia, Washington, Wisconsin, Wyoming
Repellents	Arkansas, Colorado, Connecticut, Florida, Iowa, Idaho, Illinois, Indiana, Kentucky, Maryland, Michigan, Minnesota, Mississippi, Montana, Missouri, North Carolina, North Dakota, Nebraska, New Jersey, New Mexico, New York, Ohio, Oregon, Pennsylvania, Rhode Island, Texas, Virginia, Washington, Wisconsin, Wyoming
Landscape Alteration	Alabama, Arkansas, Colorado, Connecticut, Florida, Georgia, Iowa, Idaho, Illinois, Indiana, Kentucky, Maryland, Michigan, Minnesota, Mississippi, Montana, Missouri, North Carolina, North Dakota, Nebraska, New Hampshire, New Jersey, New Mexico, Oregon, Pennsylvania, South Dakota, Texas, Virginia, Washington, Wisconsin, Wyoming
Education Programs	Arkansas, Colorado, Georgia, Iowa, Illinois, Indiana, Minnesota, Mississippi, Montana, North Dakota, Oregon, Pennsylvania, South Dakota, Texas, Virginia, Wisconsin

Note: Data for the following states was not available: AK CA DE HI LA MA NV OK UT WV

TABLE 4.3
Deer Census Techniques Practiced by States

Census Technique	States that Practice Census Technique
Harvest Surveys (mail)	Colorado, Connecticut, North Dakota, Virginia
Check Stations	Alabama, Arizona, Arkansas, Colorado, Connecticut, Iowa, Idaho, Illinois, Indiana, Kentucky, Maryland, Michigan, Minnesota, Mississippi, Montana, Missouri, North Carolina, North Dakota, Nebraska, New Jersey, New Mexico, Ohio, Oregon, Pennsylvania, South Dakota, Tennessee, Texas, Virginia, Vermont, Wisconsin
Aerial Surveys	Arizona, Arkansas, California, Connecticut, Iowa, Idaho, Illinois, Indiana, Kentucky, Maryland, Maine, Michigan, Minnesota, Montana, Missouri, North Carolina, North Dakota, Nebraska, New Jersey, New Mexico, Ohio, South Dakota, Tennessee, Texas, Virginia, Wisconsin
Thermal Imaging	Arkansas, Maryland, Mississippi, Missouri, Nebraska, Oregon, Texas
Spotlight Surveys	Alabama, Arizona, Arkansas, California, Connecticut, Florida, Idaho, Illinois, Indiana, Kentucky, Maryland, Michigan, Mississippi, Montana, Missouri, North Carolina, North Dakota, Nebraska, New Jersey, New Mexico, Oregon, Pennsylvania, South Dakota, Tennessee, Texas, Virginia, Wisconsin
Camera Monitors	Arkansas, Mississippi, Oregon, Utah
Pellet Surveys	Missouri, Pennsylvania, Virginia

Note: Data for the following states was not available: AK CA DE HI LA MA NV OK UT WV

who can assist with managing deer populations or encourage people to contact their local DNR. Many cities and municipalities will only remove dead deer carcasses or have programs to decrease population numbers.

Therefore, we concluded that the term "deer control" does not actually mean population management. "Deer control" to most people means deterring deer from their properties without harming them. We also concluded that there is a large market for repellents and deterrents. This is evident just by conducting an internet search for "deer repellent." The variety of repellents and deterrents is overwhelming.

4.5 COMMITMENT TO URBAN DEER MANAGEMENT

While conducting the research for this book, it became evident that public, academic, and agency involvement are lacking in urban deer management.

TABLE 4.4
Who Are You Going to Call? Urban Deer Management Companies and Services Provided

Company Name	Description of Services	Location of Services
Able Deer Fencing	Deer fence	Michigan
Affordable Fence TC	Deer fence	Michigan
Animal Control Solutions	**Deer control – repellents and fences**	**Nationwide**
Arbor Tree Care	Deer deterrent services	Texas
Beem Fence Company	High fence and equipment	Michigan
Benner Deer Fence Company	Heavy duty deer fence sales and installation	Michigan
Critter Control	**Preventing deer damage**	**Nationwide**
Cross Timbers Wildlife Management Services	Surveys, management, score deer	Texas
Dead Deer Removal	Dead deer removal services	Georgia
Drop-Tine Wildlife Consulting	Habitat consultant services	Pennsylvania
Habitat Solutions 360	Habitat consultant services	Michigan
Harbor Fence Company	Deer fence	Michigan
Hogarth's Pest Control Company	Deer management	Michigan
Maplehurst Fence Co.	Deer fence	Texas
Michigan Bat Control, Inc.	Repellents – sprinkler systems	Michigan
Nature's Eye Consulting	Surveys, fencing, and trapping	Texas
Nevill Fence Supply	Post and wire fence supplies	Michigan
Pro Deer Control	**Site evaluation, sharpshooting, and trapping**	**Nationwide**
San Antonio Steel Company	Deer fence	Texas
Texas Wildlife Management	Deer management, census, and trapping	Texas
The 50-year fence	Deer fence	Texas
The Management Advantage	Herd analysis, camera census, harvest recommendations, and record keeping	Texas
Town Lake Construction LLC	Electronic deer control – uses aggressive recordings	Texas
TriOrganics	Damage Control – deer repellent service	New Jersey
TX Rec Land Real Estate	Consulting services: Deer surveys/harvest recommendations	Texas
Urban Wildlife Control	Dead deer removal services	Georgia
White Buffalo Inc	**Deer population management**	**Nationwide**
Wildlife Consultants	Wildlife management plans, surveys, and harvest	Texas
Wireless Deer Fence	**Innovative deer training system**	**Nationwide**

Note: Michigan, Georgia, and Texas were used as state examples.

The key to involvement is a commitment to urban deer management. Commitment can be in the form of education and outreach, funding, and/or governmental support. In order to manage urban deer populations successfully, public education programs and state deer management plans need to be emphasized, particularly in states that have acknowledged there is an urban deer issue. By doing so, urban citizens and stakeholders will have the necessary tools to make developed and informed opinions, especially on controversial topics such as lethal population management. "All urban stakeholders concur that education is essential to the successful resolution of urban deer management problems" (Messmer et al. 1997).

For example, the Minnesota Department of Natural Resources committed itself to developing a Deer Management Plan in 2015. Plan development began in 2016, and the report was completed in 2018. The plan listed the following eight goals as part of their 2019–2028 Deer Management Plan. These goals summarize the complexity of successfully managing an urban deer population (*Minnesota Deer Management Plan 2019–2028* 2018).

1. Communication, information sharing, and public involvement: Foster trusting, respectful, and effective two-way communication between DNR and the public regarding deer management.
2. Deer stakeholder satisfaction: Consider social dimensions of deer management decisions.
3. Population management, monitoring, and research: Manage deer adaptively, considering both biological and social information in decision-making.
4. Healthy deer: Support deer herd health by monitoring and addressing disease.
5. Healthy habitat: Maintain natural wildlife habitat by protecting, enhancing, and restoring habitat and by managing for an appropriate number of deer.
6. Impact of deer on other resources: Reduce negative impacts of deer to the land, resources, and other species, including people.
7. Deer management funding: Seek sufficient funding and promote cost-effective deer management.
8. DNR deer management: Practice and ensure continuous improvement within DNR's deer management program and supporting activities.

Appendix

Online Citizen's Survey on Urban Deer Ecology and Management: Example #1

Urban Deer

1. Invitation and Explanation

Thank you for visiting our survey website. You represent one of the households in Hollywood Park or Lakeway invited to participate in a survey about urban deer management in each community. Your responses are confidential so we encourage you to answer all of the questions. In order to randomize household member representation, please select an adult occupant who will have the next birthday to answer survey questions. It should take about 25 minutes to complete this survey.

* 1. Begin by entering the four-digit access number printed above your address on the invitation card.

Access Number [ ]

White-tailed Deer in Lakeway, TX
Picture Taken By: Greg DeLong

Page 1

Urban Deer

2. Urban Deer in Your Community

2. On average, how many deer do you see in your community per week? If none, enter 0.

Number of deer:

3. On average, how many deer do you see on your property per week? If none, enter 0.

Number of deer:

4. Which of the following statements best expresses your opinion about the number of deer in your community?

- The number of deer in my community is just about right.
- There are too many deer in my community.
- There are not enough deer in my community.
- I have no opinion about the number of deer in my community.

Other (please specify)

5. Please indicate whether any of the following activities pertained to you within the last year or two. (Check all that apply)

- Fed the deer in my community.
- Advocated for the removal of deer in my community.
- Found a deer fawn.
- Rescued wounded deer.
- Allowed access to my property for deer trapping sites.
- Protested deer control measures in my community.
- Attended meetings regarding deer management in my community.
- None of the above

Other (please specify)

Page 2

Urban Deer

6. For the reasons listed below, what level of economic loss (actual or estimated dollars) have you experienced because of the deer herds in your community over the past two years? Please enter a 0 in each box if there was no economic loss in that category.

Deer/vehicle collisions

Replacement of landscape shrubs

Loss of garden crops

Health related expenses

Other losses not listed above

7. Which of the following methods have you used to control urban deer damage to your property? (Check all that apply)

- repellents
- high fences
- use plants that deer will not eat
- protect plants using tree shelters, wire cages or plastic netting
- allow deer trap sites on my property
- none of the above

Other (please specify)

Urban Deer

3. Managing Urban Deer Herds

We would like to know your opinions on various ways of managing urban deer herds.
Picture below (by Winston McInnis): bucks resting under a residents' trampoline

8. Would you favor or oppose any efforts to reduce the size of the deer herd in your community?

- strongly favor
- favor
- not sure
- oppose
- strongly oppose

9. How are urban deer currently being managed in your community?

- There is no management program that I am aware of.
- Capturing/Tranquilizing and relocating the deer
- Capturing and processing the meat
- Fertility control
- Regulated hunting with firearms
- Regulated archery hunting

Other (please specify)

Urban Deer

10. Who makes the decisions about how deer are managed in your community? (Check all the apply)

- Property Owner Association
- City Mayor
- County Judge
- Animal Rights Activists
- Residents of the community
- City Council
- Texas Parks and Wildlife Department

Other (please specify)

11. What would you consider to be evidence of a successful urban deer management program? (Check all that apply)

- Reduced number of deer/vehicle collisions
- Removal of all deer in the community
- Decreased native habitat loss
- Reduced loss of ornamental landscaping
- Improved health of the deer herd
- Lowered fertility rates in the deer herd
- Loss of a distinct browse line
- Reduced number of resident complaints
- Survival of more deer per acre
- It is impossible to successfully manage urban deer herds

Other (please specify)

12. To what extent are you informed about the methods presently used to manage deer in your community?

- a lot
- some
- not at all

Urban Deer

13. What would you consider the most effective method of delivering educational materials about urban deer management to the residents of your community? (Check all that apply)

- ê Urban Deer Management web page
- ê CD/DVD
- ê Information Booklet
- ê Recurring Television Show
- ê Question and Answer Public Meetings
- ê Workshops
- ê Educational display of information at community meeting hall
- ê Community News Letter
- ê Acquisition of reference books on deer-related topics by local public libraries
- ê Exhibits at public events
- ê Speaking engagements at homeowner associations, club or professional organization meetings
- ê Brochure

14. Many issues surface about ways to manage urban deer. How often do you consider the following issues when you think about deer management in your community.

	Never	Almost Never	Sometimes	Almost Always	Always
Safety around people	j	j	j	j	j
Method is humane	j	j	j	j	j
Deer do not suffer	j	j	j	j	j
Never fatal to deer	j	j	j	j	j
Easy to use	j	j	j	j	j
Health of the deer	j	j	j	j	j
Cost to taxpayers	j	j	j	j	j
Similar to the way nature would balance the herd	j	j	j	j	j
Allows deer to coexist with humans	j	j	j	j	j
Works quickly	j	j	j	j	j

Urban Deer

15. How acceptable are the following methods of managing urban deer herds.

	Very Unacceptable	Unacceptable	No Opinion	Acceptable	Very Acceptable
Do nothing at all.					
Allowing closely regulated hunting with firearms.					
Allowing closely regulated hunting with bow and arrow.					
Doing nothing and allowing nature to take its course.					
Using fertility control.					
Training and supervising qualified volunteers to shoot deer with firearms.					
Building fences to keep deer away from homes, roads, and parks.					
Changing the habitat so that deer have no place to live near people.					
Capturing deer and processing the meat.					
Capturing deer and moving them somewhere else.					
Hiring professional sharpshooters					

16. To what degree do you believe your opinons about deer management in your community are being considered?

- a lot
- some
- not at all

17. To what extent would you like your opinions about deer management in your community to be considered?

- a lot
- some
- not at all

Urban Deer

18. Which one of the following options to manage an overabundant urban deer herd would be the most expensive?

- Trap and relocate
- Trap and process
- Tranquilize and relocate
- Fertility Control

19. Suppose that a "Deer Action" committee was being organized to develop a plan to manage the deer herd in your community. Residents with various points of view (called stakeholders) would be expected to attend this meeting. If you attended this meeting, what type of stakeholder(s) would you represent? (Check all that apply)

- A city official
- Those who object to urban wildlife management.
- A representative of the state game agency
- Those who want to save the deer.
- Those who want to reduce the size of the deer herd.
- Disgruntled homeowner
- Those who object to hunting.
- Animal rights representative
- The media
- Those who dislike the deer for various reasons.
- A hunter

Other (please specify)

20. How is the Texas Parks and Wildlife Department involved in urban deer management? (Check all that apply)

- Issue permits for the community to legally manage their urban deer herd
- Visit communities to help form deer action committees
- Assess size of deer herds in communities
- Advise communities on how to deal with their urban deer issues
- Participate in reducing the size of the urban deer herd
- Provide funding to implement urban deer management
- Enforce the game laws associated with urban deer management
- Texas Parks and Wildlife Department is not involved in manging the deer in urban communities.
- I am not sure.

Urban Deer

4. Urban Deer: Fact or Fiction

21. Below are some general statements about urban deer. Tell us whether each statement is fact or fiction, or whether you are not sure.

	Fact	Fiction	Not sure
The size of urban habitats prevents deer herds from growing too large.			
Urban deer can carry diseases that affect humans.			
Urban deer can destroy habitat used by other animals.			
In order to survive, urban deer rely heavily on supplemental food sources.			
Urban deer can begin to produce fawns when they are six months old.			
Twins are a common result from urban deer reproduction.			
More deer per acre means the deer will be physically larger in size.			
The most common cause of death in urban deer herds is predators.			
The least common cause of death in urban deer herds is disease and starvation.			
Deer are an endangered species in Texas.			
Deer live in urban areas because they have adapted to living near people.			
People can help urban deer the most by letting nature take its course.			
Deer live in urban areas because human development has			

Urban Deer

pushed them out of their natural habitat.			
People have done more harm than good for urban deer.			
People can help urban deer the most by managing them.			
Fertility control techniques for managing urban deer are cost effective and easy to implement.			

Urban Deer

5. Recreational Preferences

22. What are your favorite outdoor activities that involve wildlife? (Check all that apply)

- Not interested in any outdoor activities involving wildlife.
- Wildlife observation or watching.
- Nature walking
- Hunting
- Fishing
- Bird watching
- Bird feeding
- Camping
- Canoeing
- Boating
- Hiking or backpacking
- Wildlife photography

Other (please specify)

Urban Deer

6. Where you live?

23. On the map below, in what part of Hollywood Park is your home located?

- I live in Lakeway. (Please scroll down to the next question.)
- A
- B
- C
- D
- E
- F

Urban Deer

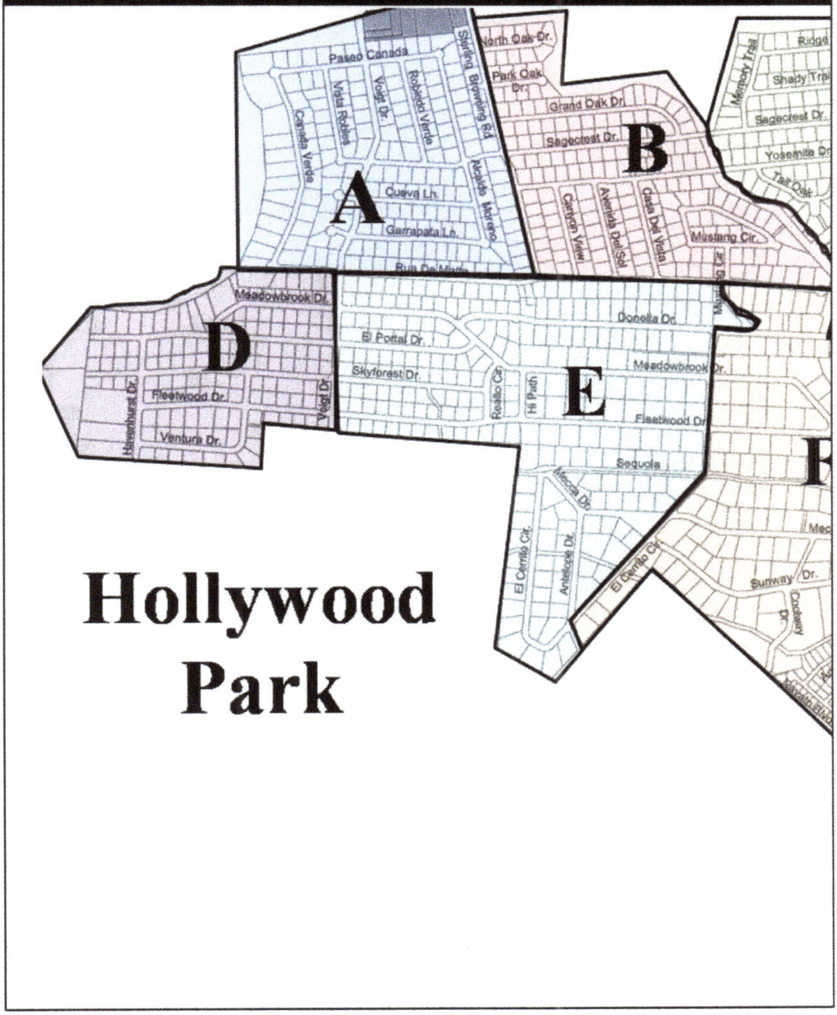

A
B
D
E
Hollywood Park
Paseo Canada
Vista Robles
Voigt Dr.
Robledo Verde
Canada Verde
Sterling Browning Rd
Cueva Ln.
Garrapata Ln.
Alcalde Moreno
North Oak Dr.
Park Oak Dr.
Grand Oak Dr.
Sagecrest Dr.
Canyon View
Avenida Del Sol
Casa Del Vista
Mustang Cir.
Memory Trail
Shady Trail
Sagecrest Dr.
Yosemite Dr.
Meadowbrook Dr.
Fleetwood Dr.
Ventura Dr.
Havenhurst Dr.
Voigt Dr.
El Portal Dr.
Skyforest Dr.
Rialto Cir.
Hi Path
Donella Dr.
Meadowbrook Dr.
Fleetwood Dr.
Sequoia
Mecca Dr.
El Cerrito Cir.
Antelope Dr.
El Cerrito Cir.
Sunway Dr.
Cooksey Dr.

Urban Deer

24. Which subdivision of Lakeway is your home located?

- I live in Hollywood Park. (Please scroll down to the next question.)
- Lakeway
- Arbolago
- Bella Strada
- Bluffs
- CV-Woods
- Casa Verde
- Cardinal Hills
- Cedar Glen
- Champions
- Cherry Mountain
- Coves
- Courtyard
- Cross Creek Villas
- Deerfield
- Estates Above Hurst Harbor
- Edgewater
- Estates of Lakeway Hills
- Enclave at Alta Vista
- Estates at Cherry Mountain
- Fairway Villas
- Flint Vallery (Flint Rock Falls)
- Flintrock (Flint Rock Falls)
- Glen Heather
- Harbor Hill
- Hurst Creek Square
- Lake Chandon
- Lake Creek Estates
- Live Oaks
- Lake Oak Estates
- Lakeway Townhomes
- Moores Treetops
- North Lakeway Village
- The Oaks
- Palomba
- Patio Homes
- Retama
- Ridge at Alta Vista
- Round Mountain Estates
- Santorini Villas
- Schooner Cove
- Schramm Ranch
- Seawind
- St Andrews
- Stoney Creek
- Sunset Park
- Terrace at the Preserve
- Terraces at Flintrock
- The Preserve at Lakeway
- Tonkawa Place
- Travis Oak Trails
- Villas at Flint Rock (Flint Rock Falls)
- Waterfront
- World of Tennis
- Yaupon Creek

Other (please specify)

25. How many people live in your household?

26. Do you own or rent your home?

- Own
- Rent

Urban Deer

27. How many years plus months have you lived at your current residence?
Example: 2 years and 6 months

Years [ ]

Months [ ]

Urban Deer

7. A Few Questions About You

Okay, you have practically finished the survey. There are just a few questions left about yourself. This information will allow us to compare your responses with others like you.

28. How many years plus months have you lived in Texas? Example: 2 years and 6 months

Years

Months

29. What year were you born?

Year 19_ _

30. Are you male or female?

- Male
- Female

31. How would you describe yourself? (Check all that apply)

- White
- Black
- Asian
- American Indian
- Of Spanish/Hispanic origin

Other (please specify)

Urban Deer

32. What is the highest level of education you have completed?

- High school graduate
- Military, trade school
- Some college
- College graduate
- Master's Degree
- Ph.D.
- Professional (law, medicine, veterinarian)

Other (please specify)

33. What is your present occupation?

34. What was your approximate total household gross income in 2007?

- Less than $30,000
- $30,000 - $39,999
- $40,000 - $49,999
- $50,000 - $59,999
- $60,000 - $69,999
- $70,000 - $79,999
- $80,000 - $89,999
- $90,000 - $99,999
- $100,000 - $109,999
- $110,000 - $119,999
- $120,000 - $129,999
- $130,000 or more

35. Is there anything else you would like to share with us? Please use this space for comments.

Urban Deer

8. End of Survey

Your participation in this survey is greatly appreciated. Consider reviewing your responses before submitting them. This survey will allow us to assess the types of information required by urban residents to more fully understand the best management practices for urban deer herds. Data collected from this survey will then be used to assist communities like Lakeway and Hollywood Park, as well as the Texas Parks and Wildlife Department, in determining the type of educational materials needed to provide relevant, current and accurate information about urban deer ecology and management for the urban resident. Thank you again for your participation.

White-tailed deer in Hollywood Park, TX

Picture Taken By: Christopher Reidy

Fair Oaks Ranch

1. Invitation and Explanation

As a resident of Fair Oaks Ranch, you are invited to participate in a survey about urban deer management. Your responses are anonymous so please answer all of the questions. All adult (18 or older) household members are encouraged to participate in this survey. It is not uncommon for household members to have different views concerning urban deer herds (white-tailed and axis). Please do not provide opportunity for others outside Fair Oaks Ranch to participate. If you have any questions about this survey, contact Clark E. Adams at cjadams@suddenlink.net.

Online Citizen's Survey on Urban Deer Ecology and Management: Example #2

Fair Oaks Ranch

2. Urban Deer in Your Commnity

1. On average, how many deer do you see in your neighborhood per week? (If none, enter 0).

Number of white-tailed deer: [ ]

Number of axis deer: [ ]

2. On average, how many deer do you see on your property per week? (If none, enter 0).

Number of white-tailed deer: [ ]

Number of axis deer: [ ]

3. Which of the following statements best expresses your opinion about the number of deer in Fair Oaks Ranch?

	Number just about right	There are too many	There are not enough	I have no opinion
White-tailed deer	○	○	○	○
Axis deer	○	○	○	○

4. Please indicate whether any of the following activities pertained to you within the last year. (Check all that apply)

- Fed the deer in my community.
- Advocated for the removal of deer in my community.
- Rescued an abandoned fawn.
- Rescued wounded deer.
- Allowed access to my property for deer trapping sites.
- Protested deer control measures in my community.
- Attended meetings regarding deer management in my community.
- None of the above

Fair Oaks Ranch

5. For the reasons listed below, what level of economic loss (actual or estimated dollars) have you experienced because of the deer herds in Fair Oaks Ranch over the past two years? Please enter a 0 in each box if there was no economic loss in that category.

Deer/vehicle collisions

Replacement of landscape shrubs

Loss of garden crops

Health related expenses

Other losses not listed above

6. Which of the following methods have you used to control deer damage to your property? (Check all that apply)

- Used repellents
- Built high fences around property
- Used plants that deer will not eat
- Protected plants using tree shelters, wire cages or plastic netting
- Tried to scare them away
- None of the above

Fair Oaks Ranch

3. Managing Urban Deer Herds

We would like to know your opinions on various ways of managing urban deer herds.
Picture below (by Winston McIinnis): bucks resting under an urban resident's trampoline

7. Would you favor or oppose any efforts to reduce the size of the deer herds in Fair Oaks Ranch?

	Strongly Oppose	Oppose	Favor	Strongly Favor	Not Sure
White-tailed deer					
Axis deer					

Fair Oaks Ranch

8. Who makes the decisions about how deer are managed in the Fair Oaks Ranch? (Check all the apply)

- City Council and mayor
- FOR Homeowners' Association
- Residents of the community
- Texas Parks and Wildlife Department
- I do not know

Other (please specify)

9. What would you consider to be evidence of a successful urban deer management program? (Check all that apply)

- Reduced number of deer/vehicle collisions
- Decreased native habitat loss
- Improved health of the deer herd
- Loss of a distinct browse line
- Survival of more deer per acre
- Removal of all deer in the community
- Reduced loss of ornamental landscaping
- Lowered fertility rates in the deer herd
- Reduced number of resident complaints
- It is impossible to successfully manage urban deer herds

Fair Oaks Ranch

10. To what extent are you informed about the methods presently used to manage deer in Fair Oaks Ranch?

- A lot
- Some
- Not at all
- Not interested

11. What would you consider the most effective method(s) of delivering educational materials about urban deer management to the Fair Oaks Ranch residents? (Check all that apply)

- Urban Deer Management web page
- CD/DVD
- Information Booklet
- Recurring Television Show
- Question and Answer Public Meetings
- Workshops
- Educational display of information at community meeting hall
- Community News Letter
- Acquisition of reference books on deer-related topics by local public libraries
- Exhibits at public events
- Speaking engagements at homeowner associations, club or professional organization meetings
- Brochure

12. Residents consider many issues when thinking about ways to manage urban deer. How often do you consider the following issues when you think about deer management in Fair Oaks Ranch?

	Never	Almost Never	Sometimes	Almost Always	Always
Deer do not suffer					
Method is humane					
Safety around people					
Easy to use					
Allows deer to coexist with humans					
Health of the deer					
Similar to the way nature would balance the herd					
Never fatal to deer					
Works quickly					
Cost to taxpayers					

Fair Oaks Ranch

13. If it were up to you, how acceptable are the following methods of managing the deer herds in Fair Oaks Ranch?

	Very Unacceptable	Unacceptable	Acceptable	Very Acceptable	No opinion
Doing nothing and allow nature to take its course.					
Using fertility control.					
Training and supervising qualified volunteers to shoot deer with firearms.					
Building fences to keep deer away from homes, roads, and parks.					
Changing the habitat so that deer have no place to live near people.					
Capturing deer and processing the meat.					
Capturing deer and moving them somewhere else.					
Hiring professional sharpshooters					

14. To what extent do you believe your opinions about deer management in your community are being considered?

- A lot
- Some
- Not at all
- Not interested

15. To what extent would you like your opinions about deer management in your community to be considered?

- A lot
- Some
- Not at all
- Not interested

Fair Oaks Ranch

16. Which one of the following methods to manage the deer herds in Fair Oaks Ranch would be the most expensive?

- Trap and relocate
- Trap and process
- Tranquilize and relocate
- Fertility Control
- I do not know

17. The Fair Oaks Ranch City Council as established A Deer Ad-Hoc Advisory Committee to develop a plan to manage the white-tailed and axis deer herds in your community. Residents with various points of view (stakeholders) would be expected to interact with the committee. What type(s) of stakeholder(s) would you represent? (Check all that apply)

- I do not know
- Those who want to save the deer.
- Those who object to hunting.
- Those who object to urban deer management.
- Those who dislike the deer for various reasons.
- A city official
- Those who want to reduce the size of the deer herd.
- A hunter
- The media
- Animal rights representative
- A representative of the state game agency
- Disgruntled homeowner

18. How is the Texas Parks and Wildlife Department involved in urban deer management? (Check all that apply)

- Issue permits for the community to legally manage their urban deer herd
- Visit communities to help form deer action committees
- Assess size of deer herds in communities
- Advise communities on how to deal with their urban deer issues
- Participate in reducing the size of the urban deer herd
- Provide funding to implement urban deer management
- Enforce the game laws associated with urban deer management
- Texas Parks and Wildlife Department is not involved in manging the deer in urban communities.
- I am not sure.

4. Urban Deer: Fact or Fiction

Fair Oaks Ranch

19. Below are some general statements about urban deer. Tell us whether each statement is fact or fiction, or whether you are not sure.

	Fact	Fiction	Not sure
Axis deer are larger than white-tailed deer and have spots.			
The size of urban habitats prevents deer herds from growing too large.			
Urban deer can carry diseases that affect humans.			
Urban deer can destroy habitat used by other animals.			
In order to survive, urban deer rely heavily on supplemental food sources.			
Urban deer can begin to produce fawns when they are six months old.			
Twins are a common result from urban deer reproduction.			
More deer per acre means the deer will be physically larger in size.			
The most common cause of death in urban deer herds is predators.			
The least common cause of death in urban deer herds is disease and starvation.			
White-tailed deer are a native Texas species.			
Axis deer are a native Texas species.			
Deer live in urban			

Fair Oaks Ranch

areas because they have adapted to living near people.			
People can help urban deer the most by letting nature take its course.			
Deer live in urban areas because human development has replaced their natural habitat.			
People have done more harm than good for urban deer.			
People can help urban deer the most by managing them.			
Fertility control techniques for managing urban deer are cost effective and easy to implement.			

5. Recreational Preferences

20. What are your favorite wildlife-related activities? (Check all that apply)

- Not interested in any outdoor activities involving wildlife.
- Wildlife observation or watching.
- Nature walking
- Hunting
- Fishing
- Bird watching
- Bird feeding
- Camping
- Canoeing
- Boating
- Hiking or backpacking
- Wildlife photography

Fair Oaks Ranch

6. Where you live?

21. What is your Foxfire address?

22. What is (are) the lot number(s) where your home is located in the Foxfire subdivision? (See subdivision plat map below)

1

2

3

Fair Oaks Ranch

23. How many people live in your household?

Number of people

24. Do you own or rent your home?

- Own
- Rent

25. How many years plus months have you lived at your current residence? Example: 2 years and 6 months

Years

Months

7. A Few Questions About You

Okay, you have practically finished the survey. There are just a few questions left about yourself. This information will allow us to compare your responses with others like you.

26. How many years plus months have you lived in Texas? Example: 2 years and 6 months

Years

Months

27. How old are you?

Age

28. Are you male or female?

- Male
- Female

29. What is the highest level of education you have completed?

- High school graduate
- Military, trade school
- Some college
- College graduate
- Master's Degree
- Ph.D.
- Professional (law, medicine, veterinarian)

30. What is your present occupation?

Fair Oaks Ranch

31. Is there anything else you would like to share with us? Please use this space for comments.

8. End of Survey

Your participation in this survey is greatly appreciated. Consider reviewing your responses before submitting them. The results of this survey will be used by the City Council and the Deer Ad-Hoc Advisory Committee to address the relevant information required by Fair Oaks Ranch residents to understand the best management practices for white-tailed and axis deer herds.

References

A Guide to Deer Management in Developed Areas of Pennsylvania. Pennsylvania Game Commission's Deer & Elk Section, Version 2.4, 2014, pp. 1–27.

"A Guide to Making an Urban Deer Management Plan." *Deer Friendly*, www.deerfriendly.com/urban-deer-management.

Adams, Clark E. *Urban Wildlife Management.* 3rd ed., CRC Press/Taylor and Francis Group, Boca Raton, FL, 2016.

Adams, C. E. "The Infrastructure for Conducting Urban Wildlife Management is Missing." *Transactions of the North American Wildlife and Natural Resources Conference* vol. 68, 2003, pp. 252–268.

Alberti, Marina, and John M. Marzluff. "Ecological Resilience in Urban Ecosystems: Linking Urban Patterns to Human and Ecological Functions." *Urban Ecosystems*, vol. 7, no. 3, 2004, pp. 241–265.

Amos, Matt, et al. "I Just Want to Count Them! Considerations When Choosing a Deer Population Monitoring Method." *Wildlife Biology*, vol. 20, no. 6, 2014, pp. 362–370.

An Evaluation of Deer Management Options. 2nd ed., Northeast Deer Technical Committee and The New England Chapter of the Wildlife Society, 2009.

Anderson, David R. "The Need to Get the Basics Right in Wildlife Field Studies." *Wildlife Society Bulletin (1973-2006)*, vol. 29, no. 4, 2001, pp. 1294–1297.

Bowker, J. M., et al. "Estimating the Economic Value of Lethal Versus Nonlethal Deer Control in Suburban Communities." *Society and Natural Resources*, vol. 16, no. 2, 2003, pp. 143–158.

Bowman, J. L. "Managing White-Tailed Deer: Exurban, Suburban, and Urban Environments." D. G. Hewitt (ed.), *Biology and Management of White-Tailed Deer.* CRC Press, Boca Raton, FL, 2011, pp. 599–620.

Branch, Geographic Products. "2010 Census Urban and Rural Classification and Urban Area Criteria." *Census Bureau QuickFacts*, United States Census Bureau, 1 September 2012, www.census.gov/geo/reference/ua/urban-rural-2010.html.

Buckland, S. T., et al. *Distance Sampling: Estimating Abundance of Biological Populations.* Chapman and Hall, London, UK, 1993.

Campbell, K. *Demographic Estimates, Home Range, and Modeling of White-Tailed Deer in the City of Fair Oaks, Texas.* Texas A&M University, Kingsville, 2014, pp. 1–132.

Cathey, J. C., et al. *Managing Overabundant White-Tailed Deer.* Texas A&M University, Texas AgriLife Extension Service, Texas AgriLife Research and Texas Parks and Wildlife Department, College Station and Austin, TX, 2010.

Chai, S. L. *Establishment of the Invasive White-Tailed Deer in Portland, Jamaica.* Jamaica Conservation and Development Trust, Kingston, Jamaica, 2003, 22pp.

Clement, Matthew J., et al. "A Method for Estimating Abundance of Mobile Populations Using Telemetry and Counts of Unmarked Animals." *Ecosphere*, vol. 6, no. 10, 2015, pp. 1–13.

Cobb, M. A. "Using Mark-Recapture Distance Sampling to Estimate Sitka Black-Tailed Deer Densities in Non-Forested Habitats of Kodiak Island, Alaska." Refuge Report no. 2014.5, Kodiak National Wildlife Refuge, U.S. Fish and Wildlife Service, Kodiak, AK, 2014.

Conover, Michael. "What Is the Urban Deer Problem and Where Did It Come From?" *Urban Deer: A Manageable Resource? Proceedings of the 1993 Symposium of the North Central Section, The Wildlife Society*, St. Louis, MO, 1995, pp. 11–18.

Cunningham, William P., and Mary Ann Cunningham. *Environmental Science: A Global Concern*. New York, NY, McGraw-Hill Higher Education, 2010.

Deckman, Dale. *North American White-Tailed Deer: Distribution and Subspecies.* Sturgeon Bay, WI, Whitetails Unlimited, Inc., 2003.

Deer Management–Citizens Task Force (Pierre, SD). City of Pierre, www.cityofpierre.org/DocumentCenter/View/128/Deer-Management---Citizens-Task-Force?bidId=.

DeNicola, Anthony J. *Managing White-Tailed Deer in Suburban Environments: A Technical Guide*. Ithaca, NY, Cornell Cooperative Extension, 2000.

DeRosa, A. J., W. Wetherbee, Brown et al. *The Urban Deer Complex: Real Tactics of the Part-Time Hunter.* Dangerous Cow Publishing, 2014.

Dewey, Tanya. "Odocoileus Virginianus (White-Tailed Deer)." Animal Diversity Web, 2003. animaldiversity.org/accounts/Odocoileus_virginianus/.

DeYoung, C. "A Population Dynamics." D. G. Hewitt (ed.), *Biology and Management of White-Tailed Deer.* CRC Press, Boca Raton, FL, 2011, pp. 147–180.

Dorney, R. S. "Biophysical and Cultural-Historic Land Classification and Mapping for Canadian Urban and Urbanizing Landscapes." E. B. Wiken and G. R. Ironside (eds.), *Ecological/Biophysical Land Classification in Urban Areas. Ecological Land Classification Series*, University of British Columbia, Vancouver, No. 3, 1978, pp. 57–71.

"Elk Facts." *Elk Network*, Rocky Mountain Elk Foundation, www.rmef.org/elk-facts/.

Franzmann, Albert W., et al. "Alaskan Moose Measurements and Weights and Measurement–Weight Relationships." *Canadian Journal of Zoology*, vol. 56, no. 2, 1978, pp. 298–306.

Frye, Bob. "A Look to the Future." *Deer Wars: Science, Tradition, and the Battle Over Managing Whitetails in Pennsylvania*, University Park, PA, Penn State University Press, 2006, p. 289.

Fuller, Matthew. *Deer Hunting in the United States: Demographics and Trends.* U.S. Fish & Wildlife Service, 2016, *Deer Hunting in the United States: Demographics and Trends.* Addendum to the 2011 National Survey of Fishing, Hunting, and Wildlife-Associated Recreation, U.S. Fish and Wildlife Service, Falls Church, VA.

Gill, R. B. *Declining Mule Deer Populations in Colorado: Reasons and Responses.* A Report to the Colorado Legislature. Colorado Division of Wildlife, 1999, pp. 1–54.

"Glossary of Environmental Terms." *Glossary of Environmental Terms*, Natural Resources Defense Council, www.nrdc.org/reference/glossary/w.asp. Retrieved on March 2, 2017.

Grimm, Annegret, et al. "Reliability of Different Mark-Recapture Methods for Population Size Estimation Tested Against Reference Population Sizes Constructed from Field Data." *PLoS ONE*, vol. 9, no. 6, 2014, pp. e98840

Grimm, Nancy B., et al. "The Changing Landscape: Ecosystem Responses to Urbanization and Pollution Across Climatic and Societal Gradients." *Frontiers in Ecology and the Environment*, vol. 6, no. 5, 2008, pp. 264–272.

Groffman, P. M., et al. "Ecological Homogenization of Urban USA." *Frontiers in Ecology and the Environment*, vol. 12, no.1, 2014, pp. 74–81.

Hansen, Lonnie, and Jeff Beringer. "Managed Hunts to Control White-Tailed Deer Populations on Urban Public Areas in Missouri." *Wildlife Society Bulletin (1973–2006)*, vol. 25, no. 2, 1997, pp. 484–487.

"Hiring a Wildlife Control Operator." Washington Department of Fish & Wildlife, wdfw.wa.gov/species-habitats/living/nuisance-wildlife/wildlife-control -operators#.

Jerozolimski, Adriano, and Carlos A. Peres. "Bringing Home the Biggest Bacon: A Cross-Site Analysis of the Structure of Hunter-Kill Profiles in Neotropical Forests." *Biological Conservation*, vol. 111, no. 3, 2003, pp. 415–425.

Johnson, G. "The Problem of Exploding Deer Populations Has No Attractive Solutions," 2015, http://biologywriter.com/on-science/articles/deerpops/.

Karns, Patrick D. "Population Distribution, Density and Trends." Charles C. Schwartz, Albert W. Franzmann, and Richard E. McCabe (ed.), *Ecology and Management of the North American Moose*. 2nd ed., Boulder, CO, University Press of Colorado, 2007, pp. 125–140.

Keller, Barbara J., et al. "A Review of Vital Rates and Cause-Specific Mortality of ElkCervus Elaphuspopulations in Eastern North America." *Mammal Review*, vol. 45, no. 3, 2015, pp. 146–159.

Key Deer Odocoileus Virginianus Clavium. U.S. Fish and Wildlife Service, https ://www.fws.gov/verobeach/msrppdfs/keydeer.pdf.

Kilpatrick, Howard J., et al. "Effects of Population Reduction on Home Ranges of Female White-Tailed Deer at High Densities." *Canadian Journal of Zoology*, vol. 79, no. 6, 2001, pp. 949–954.

Kilpatrick, Howard J., et al. "Acceptance of Deer Management Strategies by Suburban Homeowners and Bowhunters." *Journal of Wildlife Management*, vol. 71, no. 6, 2007, pp. 2095–2101.

Klimstra, W. D., et al. Key Deer Investigations Final Report: December 1967–June 1973. U.S. Fish and Wildlife Service, Big Pine Key, FL, 1974.

Kornegay, Bob. "Co-Existing with the Urban Deer Population." *Albany Herald*, 5 November 2016.

Krausman, Paul R. *Introduction to Wildlife Management: The Basics*. Upper Saddle River, NJ, Prentice Hall, 2002.

Krausman, P. R., and J. W. Cain. "Defining Wildlife and Wildlife Management". P. R. Krausman (ed.), *Wildlife Management and Conservation: Contemporary Principles and Practices*. Baltimore, MD, The Johns Hopkins University Press, 2013, pp. 1–5.

Kufeld, Roland C. "Foods Eaten by the Rocky Mountain Elk." *Journal of Range Management*, vol. 26, no. 2, 1973, p. 106.

Laliberte, J., and M. St-Laurent. “In the Wrong Place at the Wrong Time: Moose and Deer Movement Patterns Influence Wildlife-Vehicle Collision Risk.” *Accident; Analysis and Prevention*, vol. 135, 2010, pp. 105365.

Lauber, T. Bruce, and Barbara A. Knuth. “Suburban Residents’ Criteria for Evaluating Contraception and Other Deer Management Techniques.” *Human Dimensions of Wildlife*, vol. 5, no. 1, 2000, pp. 1–17.

Linnaeus, C. “Systema naturae per regna tria naturae, *secundum* classes, ordinus, genera, species, cum characteribus, differentiis, synonymis, locis.” *Editio decima, reformata. Impensis Direct. Laurentii Salvii, Holmiae*, 1758, pp. 824, http://www.biodiversitylibrary.org/item/80764page/3/mode/1up.

Mackenzie, Darryl I., et al. “Estimating Site Occupancy Rates When Detection Probabilities Are Less Than One.” *Ecology*, vol. 83, no. 8, 2002, p. 2248.

Marzluff, John M., et al. *Urban Ecology: An International Perspective on the Interaction Between Humans and Nature.* New York, NY, Springer Science Business Media, LLC, 2008.

McAninch, Jay B., ed. Urban Deer: A Manageable Resource? *Proceedings of a Symposium held at the 55th Midwest Fish and Wildlife Conference*, St. Louis, MO, 12–14 December 1993. 175pp. Symposium Chairpersons: Jay B. McAninch and Lonnie P. Hansen, 1995, pp. 1–175.

Mcclintock, Brett T., et al. “Estimating Abundance Using Mark-Resight When Sampling Is with Replacement or the Number of Marked Individuals Is Unknown.” *Biometrics*, vol. 65, no. 1, 2008, pp. 237–246.

McCullough, Dale R. *The George Reserve Deer Herd: Population Ecology of a K-Selected Species.* University of Michigan Press, Ann Arbor, MI, 1979.

McCullough, Dale R. “Lessons from the George Reserve, Michigan.” L. K. Halls (ed.), *White-Tailed Deer: Ecology and Management.* Stackpole Books, Harrisburg, PA, 1984, pp. 211–242.

McDonald, R., and P. Marcotullio. “Global Effects of Urbanization on Ecosystem Services”. J. Niemelä, J. Breuste, and T. Elmqvist (eds.), *Urban Ecology: Patterns, Processes, and Applications.* Oxford, UK, Oxford University Press, 2014, pp. 195–205.

McDonnell, M. J. and Pickett, S. T. A. “Ecosystem Structure and Function Along Urban–Rural Gradients: An Unexploited Opportunity for Ecology”. *Ecology*, vol. 71, 1990, pp. 1232–1236.

McKinney, M. L. “Urbanization, Biodiversity, and Conservation: The Impacts of Urbanization on Native Species Are Poorly Studied, but Educating a Highly Urbanized Human Population About These Impacts Can Greatly Improve Species Conservation in All Ecosystems.” *BioScience*, vol. 52, no. 10, 2002, pp. 883–890.

McKinney, Michael L. “Effects of Urbanization on Species Richness: A Review of Plants and Animals.” *Urban Ecosystems*, vol. 11, no. 2, 2008, pp. 161–176.

McKinney, Michael L. “Urbanization as a Major Cause of Biotic Homogenization.” *Biological Conservation*, vol. 127, no. 3, 2006, pp. 247–260.

McLaren, B. E., et al. “Effects of Overabundant Moose on the Newfoundland Landscape.” *Alces*, vol. 40, 2004, pp. 45–59.

Messmer, Terry A., et al. “Legal Considerations Regarding Lethal and Nonlethal Approaches to Managing Urban Deer.” *Wildlife Society Bulletin (1973–2006)*, vol. 25, no. 2, 1997a, pp. 424–429.

Messmer, Terry A., et al. "Stakeholder Acceptance of Urban Deer Management Techniques." *Wildlife Society Bulletin (1973–2006)*, vol. 25, no. 2, 1997b, pp. 360–366.

Minnesota Deer Management Plan 2019–2028. Minnesota Department of Natural Resources – Fish and Wildlife, 2018.

Montana Natural Heritage Program. "MTNHP." *Elk*, September 2007, http://fieldguide.mt.gov/speciesDetail.aspx?elcode=AMALC01010.

"Mule Deer Decline." The National Wildlife Federation, 24 November 2014, www.nwf.org/Magazines/National-Wildlife/2015/DecJan/Conservation/Mule-Deer.

Natural Resource Defense Council. Glossary of Environmental Terms, 2006, www. nrdc.org/reference/glossary/p.asp.

Nickerson, Cynthia, et al. "Major Uses of Land in the United States, 2007," EIB-89, U.S. Department of Agriculture, Economic Research Service, December 2011.

Nilon, Charles H. "Urban Biodiversity and the Importance of Management and Conservation." *Landscape and Ecological Engineering*, vol. 7, 2010, pp. 45–52.

Northern Economics, Inc. *The Value of Alaska Moose*. Prepared for the Anchorage and Alaska Soil and Water Conservation Districts, 2006, pp. 1–33.

O'Neill, R. V., and James R. Kahn. "Homo Economus as a Keystone Species." *BioScience*, vol. 50, no. 4, 2000, pp. 333–337.

O'Neill, R. V. "Is It Time to Bury the Ecosystem Concept? (with Full Military Honors, Of Course!)." *Ecology*, vol. 82, no. 12, 2001, pp. 3275–3284.

Paul, D. Curtis, et al. "Beyond Citizen Task Forces: The Future of Community-Based Deer Management." *Proceedings of the Vertebrate Pest Conference*, University of California, Davis, vol. 19, 2000, pp. 175–179.

Pickett, S. T. A., et al. "Urban Ecological Systems: Scientific Foundations and a Decade of Progress." *Journal of Environmental Management*, vol. 92, no. 3, 2011, pp. 331–362.

Pierce, B. L., R. R. Lopez, and N. J. Silvy. "Estimating animal abundance". In N. J. Silvy (ed.), *The Wildlife Techniques Manual*. 7th ed, Baltimore, MD, Johns Hopkins University Press, 2012, pp. 284–310.

Polfus, Jean L., and Paul R. Krausman. "Impacts of Residential Development on Ungulates in the Rocky Mountain West." *Wildlife Society Bulletin*, vol. 36, no. 4, 2012, pp. 647–657.

Popp, Jesse N., et al. "A Century of Elk Restoration in Eastern North America." *Restoration Ecology*, vol. 22, no. 6, 2014, pp. 723–730.

"Public Input on Deer Population Size." *Public Input on Deer Population Size – NYS Dept. of Environmental Conservation*, www.dec.ny.gov/animals/7207.html.

QDMA Staff. "QDMA's Whitetail Report 2016." QDMA's Whitetail Report 2016. "Part 2: Current Issues and Trends – Urban Deer Hunting Programs," QDMA, Bogart, GA, 2016, pp. 30–31.

Rees, W. E. "Urban Ecosystems: The Human Dimension." *Urban Ecosystems*, vol. 1, 1997, pp.63–75.

"Resource Library." *Community Deer Advisor*, deeradvisor.dnr.cornell.edu/resource-library/.

"Rutting Behavior of Moose" (U.S. National Park Service). National Parks Service, U.S. Department of the Interior, www.nps.gov/articles/aps-v5-i1-c7.htm.

Sargent, M. S., and K. S. Carter. *Managing Michigan Wildlife: A Landowners Guide*. Michigan United Conservation Clubs, East Lansing, MI, 1999, pp. 1–297.

Shaw, Bob. "Growth of Deer in Minn. Cities Divides Residents." *MPR News*, 16 January, 2011, www.mprnews.org/story/2011/01/16/urban-deer-hunting.

Siemer, W. F., T. B. Lauber, Chase, et al. "Deer/elk Management Actions in Suburban Environments: What Will Stakeholders Accept?" W. W. Shaw, L. K. Harris, and L. VanDruff (eds.), *Proceedings 4th International Urban Wildlife Symposium*. Tucson, AZ, University of Arizona, 2004, pp. 228–237.

Silvy, Nova J. *The Wildlife Techniques Manual*. Baltimore, MD, The John Hopkins University Press, 2012.

Slaughter, K., and K. Carver. *Urban Deer and Nontraditional Hunting: A Thoughtful Guide for Not-Yet Hunters*. Venison Connection, 2015.

Speller, C. F., et al. "Assessing Prehistoric Genetic Structure and Diversity of North American Elk (*Cervus elaphus*) Populations in Alberta, Canada." *Canadian Journal of Zoology*, vol. 92, no. 4, 2014, pp. 285–298.

Stewart, Chad M. "Attitudes of Urban and Suburban Residents in Indiana on Deer Management." *Wildlife Society Bulletin*, vol. 35, no. 3, 2011, pp. 316–322.

Stout, Rebecca J., et al. "Preferences of Suburban Landowners for Deer Management Techniques: A Step Towards Better Communication." *Wildlife Society Bulletin (1973–2006)*, vol. 25, no. 2, 1997, pp. 348–359.

Thomas, J. W., and D. P. Metz. *Elk of North America: Ecology and Management; Developed in Cooperation with U.S. Department of Agriculture, Forest Service*. Harrisburg, PA, Stackpole Books, 1982.

Thomas, Len, et al. "Distance Software: Design and Analysis of Distance Sampling Surveys for Estimating Population Size." *Journal of Applied Ecology*, vol. 47, no. 1, 2010, pp. 5–14.

Thompson, Jonathan R., et al. "Four Centuries of Change in Northeastern United States Forests." *PLoS ONE*, vol. 8, no. 9, 2013, pp. e72540.

Thompson, Michael J., and Robert E. Henderson. "Elk Habituation as a Credibility Challenge for Wildlife Professionals." *Wildlife Society Bulletin (1973–2006)*, vol. 26, no. 3, 1998, pp. 477–483.

Tymkiw, Elizabeth. *The Effect of White-Tailed Deer Density on Breeding Songbirds in Delaware*. University of Delaware, Newark, 2010.

"Urban Deer Population – What You Should Know." Welcome to City of Birmingham, Birmingham, MI, www.bhamgov.org/government/departments/dps/urban_deer_population.php.

Urbanek, Rachael E., et al. "Urban and Suburban Deer Management by State Wildlife-Conservation Agencies." *Wildlife Society Bulletin*, vol. 35, no. 3, 2011, pp. 310–315.

U.S. Census Bureau. "Decennial Census by Decades," April 30, 2018, www.census.gov/programs-surveys/decennial-census/decade.2010.html.

U.S. Department of the Interior, U.S. Fish and Wildlife Service, and U.S. Department of Commerce, U.S. Census Bureau. "2016 National Survey of Fishing, Hunting, and Wildlife-Associated Recreation," 2016.

VanDruff, L. W., et al. "Management of Urban Wildlife." Theodore A. Bookhout (ed.), *Research and Management Techniques for Wildlife and Habitats.* 5th ed., Bethesda, MD, Wildlife Soc. (1999), 1996.
Vercauteren, Kurt C., et al. "Regulated Commercial Harvest to Manage Overabundant White-Tailed Deer: An Idea to Consider?" *Wildlife Society Bulletin*, vol. 35, no. 3, 2011, pp. 185–194.
Vermeire, Lance, M. Wallace, and R. Mitchell. "Elk Populations in the Western United States and Canadian Provinces: Discover How Game Agencies Count Elk, Why, and the Accuracy of the Estimates." *Rangelands*, vol. 26, no. 5, 2004, pp. 29–33.
Von Drehle, David. "America's Pest Problem: It's Time to Cull the Herd." *TIME Magazine*, 9 December 2013.
Wakeling, B. F., et al. *Mule Deer and Movement Barriers.* Mule Deer Working Group, Western Association of Fish and Wildlife Agencies, USA, 2015.
Waller, Donald M., and William S. Alverson. "The White-Tailed Deer: A Keystone Herbivore." *Wildlife Society Bulletin (1973–2006)*, vol. 25, no. 2, 1997, pp. 217–226.
White, Gary C., and Kenneth P. Burnham. "Program MARK: Survival Estimation from Populations of Marked Animals." *Bird Study*, vol. 46, no. supl, 1999, pp. 120–138.
"Wildlife Control Operators List." *South Carolina DNR*, South Carolina DNR, www.dnr.sc.gov/wildlife/docs/nwco.pdf.
Wildlife Habitat Management Institute. "American Elk (*Cervus elaphus*)." Fish and Wildlife Habitat Management Leaflet. USDA. Natural Resources Conservation Service, no. 11, 1999, 8pp, https://www.nrcs.usda.gov/Internet/FSE_DOCUMENTS/nrcs143_010000.pdf.
Worm, Boris, and Robert T. Paine. "Humans as a Hyperkeystone Species." *Trends in Ecology and Evolution*, vol. 31, no. 8, 2016, pp. 600–607.
Wright, Richard T., and Dorothy Boorse. *Environmental Science: Toward a Sustainable Future, 11th ed.* Boston, MA, Benjamin Cummings, 2011.

Index

For Product Safety Concerns and Information please contact our EU representative GPSR@taylorandfrancis.com
Taylor & Francis Verlag GmbH, Kaufingerstraße 24, 80331 München, Germany

www.ingramcontent.com/pod-product-compliance
Ingram Content Group UK Ltd.
Pitfield, Milton Keynes, MK11 3LW, UK
UKHW022211190926
475702UK00003B/303
* 9 7 8 0 3 6 7 4 8 1 6 2 9 *